CRAM SESSION IN

Functional Anatomy

A Handbook for Students & Clinicians

CRAM SESSION IN
Functional Anatomy
A Handbook for Students & Clinicians

SCOTT BENJAMIN, PT, DSCPT
Exclusive Physical Therapy
Lansing, Michigan

Roy H. Bechtel, PT, PhD
University of Maryland School of Medicine
Baltimore, Maryland

Vincent M. Conroy, PT, DScPT
University of Maryland School of Medicine
Baltimore, Maryland

Routledge
Taylor & Francis Group

NEW YORK AND LONDON

First published in 2011 by SLACK Incorporated

Published in 2024 by Routledge
605 Third Avenue, New York, NY 10158

and by Routledge
4 Park Square, Milton Park, Abingdon, Oxon, OX14 4RN

Routledge is an imprint of the Taylor & Francis Group, an informa business

© 2011 Taylor & Francis Group

Library of Congress Cataloging-in-Publication Data

Benjamin, Scott.
  Cram session in functional anatomy: a handbook for students and clinicians / Scott Benjamin, Roy H. Bechtel, Vincent M. Conroy.
    p. ; cm.
  Includes bibliographical references and index.
  ISBN 9781556429361 (alk. paper)
1. Musculoskeletal system--Anatomy. I. Bechtel, Roy H. II. Conroy, Vincent M. III. Title.
  [DNLM: 1. Anatomy--Handbooks. QS 39]
  QM100.B46 2011

  611'.7--dc22

                                    2010026396

ISBN: 9781556429361 (pbk)
ISBN: 9781003523383 (ebk)

DOI: 10.4324/9781003523383

CONTENTS

ACKNOWLEDGMENTS

As with any appreciation statement, it is hard to determine who to thank, but first and foremost I would like to say "thank you" to Brien Cummings of SLACK Incorporated. Without his support and guidance, this project would have not occurred and we appreciate the opportunity to work with him on this manuscript and more in the future. I would like to thank the Lord above for giving me the desire to write and for teaching me how important it is to give back to others in any area, whether practical or didactic. Thank you to Drs. Roy Bechtel and Vinnie Conroy for wanting to undertake this project around their teaching and research schedules; one more thing added to their plate hopefully did not tip over the stack! Most importantly, thank you to my wife, who has endured my wanting to write for years and has encouraged me to finish projects. She is my biggest fan through rejection and my best friend along this curvy road of life; I thank you from my heart. To the students who will read this book—never stop learning and keep persevering toward your goals! —SB

This book is for Carol! —RB

A special additional thank you goes out to Kenda Robinson and Karen Townsend for their help with all the photos, and to Lora Benjamin for her great eye in taking all the still photos. We sincerely appreciate all the help on this project. —SB, RB, VC

ABOUT THE AUTHORS

Scott Benjamin, PT, DScPT, received his undergraduate training from the Michigan Technological University in 1982 and his BS in Physical Therapy degree from the University of Illinois at Chicago in 1989. His graduate work was completed at the University of Maryland and he is the coauthor, with Dr. Bechtel, of one of the first texts on rehabilitation after artificial disc replacement in the country. He has authored papers on total disc arthroplasty, lateral epicondylitis, clinical modalities, isokinetic testing, and aquatic therapy. Scott teaches continuing education courses nationally with Dr. Bechtel. His areas of interest are biomechanics of the spine and sacroiliac joint, manual therapy, and rehabilitation.

Roy H. Bechtel, PT, PhD, graduated from the University of Maryland with a BS in Physical Therapy in 1979. He received an MS in Physical Therapy from New York University in 1981 and a PhD in Biomechanics from the University of Maryland in 1998. He teaches in the Department of Physical Therapy and Rehabilitation Science in the School of Medicine at the University of Maryland in Baltimore, and has conducted continuing education courses nationally and internationally. His research interests are in manual physical therapy, assessment and treatment of pain of spinal origin, and biomechanical modeling of forces applied to spine and sacroiliac joints. Recently, Dr. Bechtel published a paper on the tolerance for isokinetic testing pre- and post-lumbar fusion. He is the coauthor, with Dr. Benjamin, of a book on postsurgical rehabilitation after artificial disc replacement.

Vincent M. Conroy, PT, DScPT, is a two-time graduate of the University of Maryland, receiving a BS in Physical Therapy in 1990 and his DScPT with an emphasis on Basic Science in 2005. Dr. Conroy currently holds the rank of Assistant Professor at the University of Maryland School of Medicine, Department of Physical Therapy and Rehabilitation Science (PTRS). His primary duties include instruction of Human Anatomy, Biomechanics, and Musculoskeletal Screening Evaluation and Treatment. He also provides oversight as the Clinical Director of the Department of PTRS's Service Learning Center (SLC). The SLC provides pro-bono physical therapy intervention to individuals who do not have the benefit of health insurance. Dr. Conroy has lectured both regionally and nationally on topics related to the SLC and clinical anatomical reasoning.

PREFACE

This book is written for the practicing clinician or student. Designed as a quick reference, the project provides a general overview of the muscular systems, with a focus on the muscles commonly encountered during rehabilitation. The book brings to light current best evidence regarding muscle function and biomechanics, and provides clinical pearls that should be invaluable for dealing with specific clinical situations.

FOREWORD

This is the second book by Roy Bechtel and Scott Benjamin for which I have had the privilege of writing a foreword, and in keeping with the design of this book I will be concise and accurate in describing the value of it. I have known Roy and Scott for many years as excellent clinicians and clinical educators (this despite Roy's PhD) and this book reflects their ability in this area. Although the book is principally aimed at pre-professional students and new graduates in physical therapy, it is extremely valuable to those involved in movement rehabilitation who would like to have a short but useful review of descriptive and functional anatomy, or have a need for a simple text for their students. By combining the more vital facts of the subject with a clear method of information organization, the authors have managed to simplify a very complicated subject without making it simplistic. The relevant anatomy is found in tabulated form with clear diagrams that allow the reader to quickly review before going on to the meat of the subject—rehabilitation. This takes the form of using functional anatomy and biomechanics together with a wealth of clinical experience and research and other forms of evidence to guide the therapist to a rational and reasoned approach to the treatment of musculoskeletal disorders. The text and the pictures make it very clear what the authors are describing, which in turn makes it easy for even the newest graduate or student to follow their approach, especially after reviewing the anatomy from their new clinical perspective.

I have no reservations about recommending this valuable book to any therapist of any level as a guide to specific exercise prescription, a reference for anatomy and biomechanics, and as an instructional text for students.

James T. S. Meadows, BScPT, FCAMT
Vice President of Curriculum for the
North American Institute of Orthopedic
Manual Therapy

INTRODUCTION

Our students, whether they are physical therapy students or medical students, have made us appreciate the value of simplicity. Keeping things simple makes even complex topics seem understandable. The challenge of dealing with a complex topic like the neural control of muscle function is to transmit the relatively basic understanding that we currently have of this important subject without giving the impression that we have presented a complete picture of muscle function. Research on neuromotor control during functional activities is in its infancy, and most models of neural control are based on incomplete and/or highly circumscribed data sets. It is necessary, therefore, to rely on clinical experience to fill the gaps in our present understanding of muscle function. We believe that this approach, combining research data, where available, with clinical expertise and a considerable regard for patient variability, fulfills the criteria for the practice of evidence-based medicine, as outlined by Straus et al.[1] In this book, we will try to clearly distinguish between research evidence and clinical evidence to allow the best synthesis of the material for patients and for providers.

The book is divided into 3 main sections: The Upper Limbs, The Lower Limbs, and The Spine. Chapters will give a brief overview of the region, relating current best evidence to understanding the anatomical relationships and clinical correlations. They will allow the reader to access a more in-depth appreciation for regional bio- and pathomechanics. Within each section, important muscles will be discussed in a readily accessible fashion, according to the following schema: function, origin, pathophysiology, clinical pearls, and illustrations of the muscle in vivo and also internally with muscle forces demonstrated.

Reference

1. Straus SE, Richardson WS, Glasziou P, Haynes BR. *Evidence-Based Medicine: How to Practice and Teach EBM.* 3rd ed. Philadelphia, PA: Churchill Livingstone; 2005.

SECTION I

The Upper Limbs

Benjamin S, Bechtel RH, Conroy VM.
Cram Session in Functional Anatomy:
A Handbook for Students & Clinicians (pp. 1-44)
© 2011 Taylor & Francis Group.

1

THE TRAPEZIUS AND MUSCLES THAT AFFECT THE NECK

You are in the middle of a big project, typing away madly under those deadlines and time constraints. You can feel the tension at the top of your neck and also in your shoulders. You can feel the tightness increasing by the minute and your shoulder starts to lift up toward your ear. This is the upper trapezius (along with levator scapula and the scalenes) cranking up the tension on your neck and shoulders. This section will discuss the upper trapezius and its importance.

Function of the Upper Trapezius

The trapezius is a large triangular muscle that spans from the neck to the bottom of the thoracic spine on both sides.[1] Each trapezius is divided into 3 parts, named for their location on the back as upper, middle, and lower trapezius (Figure 1-1). All the parts are innervated by cranial nerve 11, the spinal accessory. The upper trapezius is the most superficial of the back muscles[2] and acts on the neck as well as the shoulder girdle (Figure 1-2).

When a patient presents with pain along the top of the shoulder and neck with accompanying shoulder motion dyskinesis, the trapezius may be involved.[3] Specifically, the lower trapezius fibers may be overworking and the upper fibers may be inhibited and weak.[2] Normal daily activities are a series of timed events that the muscles need to coordinate so that movement is allowed and stability is maintained. The trapezius plays an integral part in the coordinated movements of the shoulder and arm.[4] Overtraining, undertraining, or neglect of any of the muscles that surround the shoulder can disrupt the balance of shoulder function.[5,6] This can lead to shoulder motion dyskinesis, along with a reduction in the person's functional level and pain caused by poor scapular stability during movement. The upper and middle trapezius fibers are responsible for pain referral patterns that can extend from the head and neck to the scapula.[7-10]

Patients who work with computers rely on the upper trapezius for mouse activation, typing, and any activity that requires reaching, like putting papers in a pile.[11,12] Excessive trapezius recruitment can lead to headaches, tightness, and soreness that can refer downward to your shoulder blade or up into your neck.[13-15] The trapezius soreness can come from either the upper, middle, or lower fibers and what is interesting is that certain

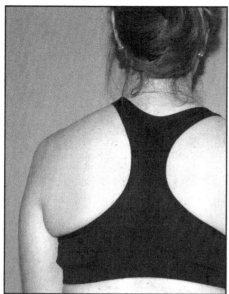

Figure 1-1A. Left trapezius relaxed in sitting.

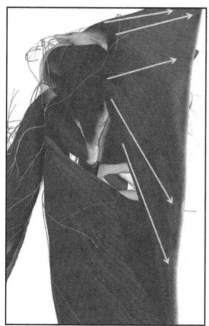

Figure 1-1B. Schematic of the left trapezius with the muscle forces shown by the arrows for the upper, middle, and lower trapezius. (Reprinted with permission from Primal Pictures, 2009.)

Figure 1-2A. The right upper trapezius muscle contracted in the sitting position.

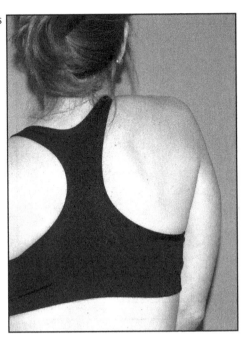

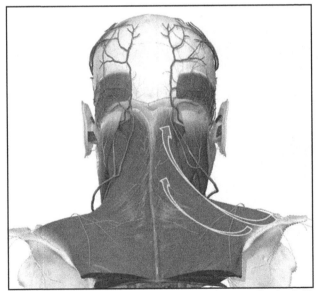

Figure 1-2B. The right upper trapezius forces as noted by the arrows. (Reprinted with permission from Primal Pictures, 2009.)

functional activities can cause pain to be elicited from specific trapezius fibers in a unique fashion.[15,16] Pain referral patterns to the head, neck, and back can also be associated with compression of perforating vasculature and cutaneous nerves that target the skin and superficial fascia. Examples in the head and neck region are the greater occipital nerve, lesser occipital nerve, third occipital nerve, and occipital artery. Segmental neurovascular bundles are named accordingly, in descending fashion, after the third cervical level.[17-19] Driving with one hand on top of the steering wheel calls into play the lower trapezius. Characteristically, lifting activates the upper trapezius, and a person whose job entails repetitive lifting may present with one shoulder higher than the other as a postural habit. An inhibited middle trapezius can contribute to a rounded shoulder posture.[10,15]

Origin and Insertion

Origin of the Trapezius Muscle	Insertion of the Trapezius Muscle
Originate from the external occipital protuberance and the superior nuchal line of the occipital bone as well as the ligamentum nuchae and from the ligamentum nuchae and supraspinous ligament in the cervical and thoracic spine levels.[1,19,20]	The upper fibers attach to the lateral third of the clavicle at its posterior aspect of the superior surface.[19,20] The middle portion attaches to the posterior border of the acromion and the cranial lip. The inferior fibers travel to the lateral portion of the scapulae to attach to the base of the scapular spine.[1,19,20]

ACTION OF THE UPPER TRAPEZIUS
If the scapula is fixed, the upper trapezius will cause ipsilateral side flexion and contralateral rotation, similar to the sternocleidomastoid muscle, which produces torticollis or wry neck.[2] When both upper trapezius muscles contract they produce head and neck extension.[19,21] The upper fibers will aid in scapular upward rotation that occurs with abduction. The lower fibers will assist the middle and upper fibers with upward rotation but will also cause the scapula to rotate in a downward fashion.[20] All the parts of the trapezius cause scapular adduction, and this muscle makes an important contribution to scapular stability.

Pathology of Common Injuries

The trapezius muscle is subject to injury just like any other muscle but its affects can also affect other parts of the body (ie, neck, shoulder, mid-back).[2] Fernandez-de-Las-Penas et al[6] discovered that the upper trapezius fibers are associated with **headaches** and also with tension that can refer pain to the neck. Since the trapezius is innervated by the cranial accessory nerve (XI), a person who presents with pain in his or her neck can have symptoms arise from the central nervous system as well as the peripheral nervous system.[22] The Rosendal study indicated that if upper trapezius is not strengthened to a proper baseline level, headaches can be brought on by work-related tasks, and this is an indication of the need to start an exercise program that would benefit the trapezius.[22,23] With all these variables, work-related trapezius pain will vary, depending upon the reaching height; angles of shoulder, elbow, and wrist; and how much muscle activity is required within the trapezius; thus the level of pain can be quite variable.[14]

The upper fibers of the trapezius have been most closely linked to **cervicogenic and tension-related headaches.**[7,14,24] Studies have shown that the more this muscle is active, the probability that a headache will occur increases.[14,25] Biondi et al[26] reported cranial nerve and central nervous system tract involvement, plus an inflammatory cascade involving the spinal accessory nerve (XI) as potential mechanisms for cervicogenic headaches. The net effect may be to ramp up not only muscle tension, but also the afferent input from muscles like the trapezius, causing tightness in the neck and occiput.[26,27] Additionally, when a person experiences cervicogenic or tension headaches, the associated fascia is also stretched, which applies mechanical strain to an area that is already overstimulated and oversensitized.[7,16,28] The muscles that are usually affected in tension headaches are the upper trapezius, the sternocleidomastoid, and the scalene muscles. Humphreys et al[29] showed that myodural connections exist between the ligamentum nuchae and the cervical dura mater, and Hack and Hallgren[30] showed that **chronic headache** may be related to a similar connection between rectus capitis minor and the cervical dura mater. All must be looked at and can be treated with various manual and pain-relieving modalities to relieve some of the pain within the muscular systems.[31,32]

Clinical Pearls: Trapezius

- Aids in shoulder stability by stabilizing the scapulae

- Aids in shoulder motion and supports the rotator cuff

- Has a direct effect on tension headaches

- Has a direct relationship with cervicogenic headaches

- Linked to changes in the mechanics of the shoulder when surgery or an injury has occurred

- If adversely affected, there is a link between shoulder impingement syndrome and the function of the trapezius muscle

References

1. Warfel J. *The Extremities: Muscles and Motor Points.* Philadelphia, PA: Lea and Febiger; 1985.

2. Dutton M. *Orthopaedic Examination, Evaluation, and Intervention.* New York, NY: McGraw-Hill: 2008.
3. Sahrmann S. *Diagnosis and Treatment of Movement Impairment Syndromes.* New York, NY: Mosby; 2001.
4. Hughes AM, Freeman CT, Burridge JH, et al. Shoulder and elbow muscle activity during fully supported trajectory tracking in neurologically intact older people. *J Electromyogr Kinesiol.* 2008;19:1025-1034.
5. Kolber MJ, Beekhuizen KS, Cheng MS, Hellman MA. Shoulder joint and muscle characteristics in the recreational weight training population. *J Strength Cond Res.* 2009;23(1):148-157.
6. Alexander C, Miley R, Stynes S, Harrison PJ. Differential control of the scapulothoracic muscles in humans. *J Physiol.* 2007;580(Pt.3):777-786.
7. Fernandez-de-Las-Penas CH, Ge Y, et al. Referred pain from trapezius muscle trigger points shares similar characteristics with chronic tension type headache. *Eur J Pain.* 2007;11(4):475-482.
8. Thorn S, Sogaard K, et al. Trapezius muscle rest time during standardised computer work—a comparison of female computer users with and without self-reported neck/shoulder complaints. *J Electromyogr Kinesiol.* 2007;17(4):420-427.
9. Mathiassen SE, Nordander C, et al. Task-based estimation of mechanical job exposure in occupational groups. *Scand J Work Environ Health.* 2005;31(2):138-151.
10. Zennaro D, Laubli T, et al. Motor unit identification in two neighboring recording positions of the human trapezius muscle during prolonged computer work. *Eur J Appl Physiol.* 2003;89(6):526-535.
11. Thorn S, Forsman M, et al. A comparison of muscular activity during single and double mouse clicks. *Eur J Appl Physiol.* 2005;94(1-2):158-167.
12. Szeto GP, Straker LM, et al. A comparison of symptomatic and asymptomatic office workers performing monotonous keyboard work—1: neck and shoulder muscle recruitment patterns. *Man Ther.* 2005;10(4):270-280.
13. Johnson EG, Godges JJ, et al. Disability self-assessment and upper quarter muscle balance between female dental hygienists and non-dental hygienists. *J Dent Hyg.* 2003;77(4):217-223.
14. Westgaard RH. Muscle activity as a releasing factor for pain in the shoulder and neck. *Cephalalgia.* 1999;19(Suppl 25):1-8.
15. Simons DG, Travell JG. Myofascial origins of low back pain. 3. Pelvic and lower extremity muscles. *Postgrad Med.* 1983;73(2):99-105,108.
16. Fernandez-de-Las-Penas C, Simons D, et al. The role of myofascial trigger points in musculoskeletal pain syndromes of the head and neck. *Curr Pain Headache Rep.* 2007;11(5):365-372.
17. Bogduk N, Marsland A. On the concept of third occipital headache. *J Neurol Neurosurg Psychiatry.* 1986;49:775-780.
18. Ducic I, Moriarty M, Al-Attar A. Anatomical variations of the occipital nerves: Implications for the treatment of chronic headaches. *Plastic and Reconstructive Surgery.* 2009;123(3):859-863.
19. Standring S. *Gray's Anatomy: The Anatomical Basis of Clinical Practice.* 40th ed. New York, NY: Churchill Livingstone; 2008.
20. Clemente CD. *Gray's Anatomy.* 30th ed. Philadelphia, PA: Lea and Fabinger; 1985.
21. Kendall F, McCreary E, Provance P, Rodgers M, Romani W. *Muscles: Testing and Function With Posture and Pain.* 5th ed. New York, NY: Lippincott Williams & Wilkins; 2005.

22. Rosendal L, Kristiansen J, et al. Increased levels of interstitial potassium but normal levels of muscle IL-6 and LDH in patients with trapezius myalgia. *Pain.* 2005;119(1-3): 201-209.

23. Falla D, Bilenkij G, et al. Patients with chronic neck pain demonstrate altered patterns of muscle activation during performance of a functional upper limb task. *Spine.* 2004;29(13):1436-1440.

24. Moore MK. Upper crossed syndrome and its relationship to cervicogenic headache. *J Manipulative Physiol Ther.* 2004;27(6):414-420.

25. Sandrini G, Antonaci F, et al. Comparative study with EMG, pressure algometry and manual palpation in tension-type headache and migraine. *Cephalalgia.* 1994;14(6):451-457; discussion 394-455.

26. Biondi DM. Cervicogenic headache: diagnostic evaluation and treatment strategies. *Curr Pain Headache Rep.* 2001;5(4):361-368.

27. Hull G, Barrett C, et al. Further clinical clarification of the muscle dysfunction in cervical headache. Cephalalgia. 1999;19(3):179-185.

28. Masi AT, Hannon JC. Human resting muscle tone (HRMT): narrative introduction and modern concepts. *J Bodyw Mov Ther.* 2008;12(4):320-332.

29. Humphreys BK, Kenin S, Hubbard BB, Cramer GD. Investigation of connective tissue attachments to the cervical spinal dura mater. *Clin Anat.* 2003;16(2):152-159.

30. Hack GD, Hallgren RC. Chronic headache relief after section of suboccipital muscle dural connections: a case report. *Headache.* 2004;44(1):84-89.

31. Kim HS, Chung SC, et al. Pain-pressure threshold in the head and neck region of episodic tension-type headache patients. *J Orofac Pain.* 1995;9(4): 357-364.

32. King TI. The use of electromyographic biofeedback in treating a client with tension headaches. *Am J Occup Ther.* 1992;46(9):839-842.

2

THE DELTOID

As you walk, work out, or ride your bike up Mont Ventoux, your deltoids and rotator cuff muscles are helping to stabilize the shoulder. The deltoid is a large muscle that has a large impact on the mechanics of the shoulder and is a prime mover of the glenohumeral joint. This chapter will outline the functions of the deltoid, discuss issues of muscular control, and also provide clinical pearls for this muscle.

Function of the Deltoid

The deltoid is a very large muscle with 3 heads (anterior, middle, and posterior) that makes up about 20% of all shoulder muscle bulk (Figures 2-1, 2-2, and 2-3).[1] The deltoid muscle and the rotator cuff together control the majority of arm motion at the shoulder joint.[2-4] The deltoid is active during arm elevation and also during many rehabilitation exercises.[4] However, if the deltoid works alone, without its supporting cast of the rotator cuff muscles and the long head of biceps brachii, there will not be humeral head depression during movement, which can cause impingement to occur between the humeral head and the coracoacromial ligament. If this situation persists, it can lead to further shoulder pathology.[2,5]

When persons use their shoulder and they do not have their upper extremity down on a stable base (closed kinetic chain position), they may activate more of the deltoid to support their shoulder against gravity.[6] The deltoid, like most muscles, will become more or less active depending upon the demands placed on it.[7] Michiels et al[8] used EMG to show that the deltoid muscle contributes to many different stability patterns, depending on the motions involved. The 3 muscle systems together (deltoid, long head of the biceps, and rotator cuff) are important to control shoulder motions and stability, but a look at the cross-sectional area of each of these muscles can give an understanding of their relative importance to normal shoulder joint mechanics and daily function. For example, if a person has his or her arm abducted, in scaption, or in a flexed position, the deltoid is providing a vast amount of stability to the shoulder along with the supportive rotator cuff and scapulothoracic muscles. Besides the dynamic support of the active shoulder muscles, stability of the shoulder does rely on the passive structures (ie, the ligamentous and capsular system), which we will touch on in the next section, but with the deltoid, rotator cuff, and the long head of the bicep, dynamic control of the glenohumeral joint can occur.[9]

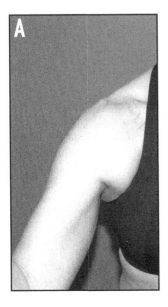

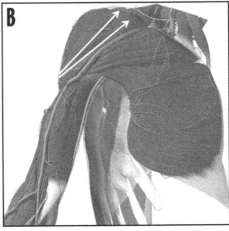

Figure 2-1. (A) The right anterior deltoid in vivo. (B) Schematic of the right anterior deltoid. (Reprinted with permission from Primal Pictures, 2009.)

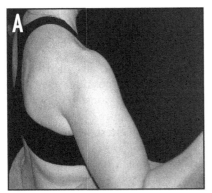

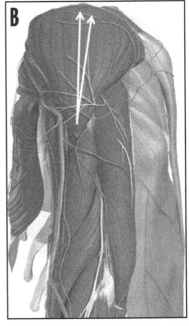

Figure 2-2. (A) Middle deltoid muscle in vivo. (B) Schematic of the middle deltoid muscle with the arrows showing the line of muscle force. (Reprinted with permission from Primal Pictures, 2009.)

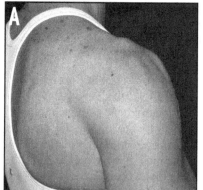

Figure 2-3. (A) The posterior deltoid muscle in vivo. (B) Schematic of the posterior deltoid with muscle forces from the posterior portion of the muscle fibers. (Reprinted with permission from Primal Pictures, 2009.)

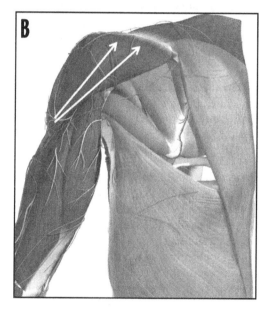

Passive Inherent Stability

Despite the important role of the dynamic stabilizers, they are only effective as long as the passive stabilizers are intact.[9-12] This stability comes from the inherent passive structures that will support the shoulder, such as the glenoid labrum (which is a fibrocartilaginous ring around the rim of the glenoid fossa). The glenoid labrum deepens the fossa and serves as an attachment point for the long head of biceps brachii. With the mobile characteristics of the glenohumeral joint, there is a requirement for dynamic stability from the muscles that surround the shoulder. Working together, passive and active structures will provide a mechanism that can control osteokinematic movements between the glenoid and the head of the humerus throughout all positions and ranges of movement.

The shoulder requires multidirectional dynamic stability during sporting and daily life activities. The passive stabilizers that supply this multidirectional stability are the ligaments that control bone-to-bone motion.[11] This would include the shoulder capsule and in particular the superior, middle, and the inferior glenohumeral ligaments, which are situated within the capsule and are considered some of the main passive stabilizers of the glenohumeral joint.[9,11] Studies have shown that when you lose the passive stabilizers (eg, glenoid labrum) you can lose 20% or more of the stability of the shoulder.[13] Any lack of passive stability will place a greater demand on the muscular system to control and provide stability for the shoulder.[13] As a person ages, his or her shoulder tissues will do the same and there could be a loss of passive stability, which will put undue stress upon the dynamic muscular stabilizers. At the same time, muscular strength is declining. This may suggest that we need to focus the rehabilitation to accommodate for these changes in older patients and emphasize working on the large muscular system, balance, and motor control activities.

The physiologic cross-sectional area of the shoulder muscles (cross-sectional area divided by the fiber length) was determined by Bassett.[1] He concluded that the deltoid accounts for 22.6% of the shoulder muscle mass, the rotator cuff accounts for 19%, and the long head of the bicep only accounts for 3.9%.[1,9] An injury to the long head of the biceps will not cause a major loss of dynamic stability, however, due to its mechanical properties, will deprive the shoulder of an important mechanism of resisting humeral superior translation, which will undoubtedly avoid the chance of impingement. Injury to the long biceps will require the rotator cuff to become more active, since the deltoid cannot provide significant depressive force on the humerus. In addition, because of its attachment to the glenoid labrum, injury to the long head of the biceps brachii tendon can tear the labrum, reducing passive stability and possibly resulting in a SLAP (superior labrum from anterior to posterior) lesion[14] and an unstable shoulder. When the deltoid is torn or injured, shoulder motion and function are significantly hampered and the rotator cuff must work harder to maintain normal shoulder biomechanics during motion.[9] The deltoid, along with its supporting cast of shoulder and scapular stabilizing muscles, will help create the 2:1 ratio of motion that occurs between the glenohumeral joint and the scapulothoracic articulation.[9,15] We now know that this ratio, while constant overall, is quite variable during different stages of movement, and the deltoid plays a vital role along with rotator cuff, serratus anterior, and the long head of the biceps to control this action.[9]

Origin and Insertion

Origin of the Deltoid Muscle	Insertion of the Deltoid Muscle
The deltoid is a large and very thick triangular muscle that arises from the lateral third of the clavicle (anterior head), superior acromion surface (middle head), and the caudal lip of spine of the scapulae (posterior head).	The 3 heads blend together and converge to insert on the deltoid tuberosity of the humerus.[10,11,16,17]

ACTION OF THE DELTOID MUSCLE
The 3 heads of the deltoid, working with rotator cuff muscles, help to compress the humeral head into the glenoid cavity to provide this dynamic stability, even as the deltoid supplies the power to produce a range of shoulder movements such as abduction, flexion, extension, and rotation (see Figures 2-1, 2-2, and 2-3).[10]

Pathology of Common Injuries

The deltoid muscle is less prone to injury in the shoulder than the rotator cuff or the long head of the biceps.[9] **The deltoid muscle is subject to bursitis at its insertion point**[10] and this can lead to pain when the deltoid compresses the bursa with abducted or flexed motions. When a person has some sort of shoulder surgery, either open or arthroscopic, the deltoid muscle will also be intimately involved. In these cases, there could be complications that result in dysfunction of the deltoid muscle and lead to a longer recovery or more extensive rehabilitation process.[18] The deltoid can also become **fibrotic**[19] as a result of atrophy or radiation treatment for cancers associated with the head or neck region. This condition is rare, but can occur, and may need surgical treatment to correct.

When a person is playing sports, whether he or she is contact oriented or not, the shoulder muscles are being used. The order at which the activity will give rise to muscle-coordinated movement depends on the sport that is being played. Even though the deltoid muscle is not the most commonly injured muscle, it is subject to trauma and injury. For example, tennis players, as they serve, will use many shoulder muscles, including the anterior and middle heads of the deltoid.[5] In cases where the deltoid is injured, the tennis serve could be compromised and thus the tennis player will rely more heavily on the long head of the bicep and the rotator cuff muscles.[5,20] When that occurs, the mechanics of the shoulder could be compromised, and the person may experience altered mechanics of motion that could result in serve impingement or even a tear of the muscle.

Another injury that may occur to the deltoid can occur with a **direct trauma** to the proximal humerus, as in hockey or other contact sporting activities, which could result in a **contusion** or even **subluxation** and resulting injury to the axillary nerve that supplies the deltoid muscle.[21-23] Besides direct trauma from contact sports, patients who dislocate their shoulder can experience trauma to their axillary nerve,[22] which can result in a traction force applied to the nerve to reduce its outflow to the deltoid muscle leading to weakness or paralysis. The deltoid may also be injured along with a rotator cuff tear depending on the severity of the injury.[24] To correct this problem, surgery is typically performed. Leaving the deltoid attachment undisturbed will result in shorter recovery time and faster healing and recovery of function.[25]

Lastly, all of the tendons around the shoulder have been reported to be susceptible to tendinitis, which can lead to changes in function. The deltoid is no exception to this, but the problem of **tendinitis and/or calcifications** here is relatively rare.[26]

Clinical Pearls: Deltoid Muscle

- Works in concert during shoulder biomechanics and osteokinematic motion

- Aids the rotator cuff in shoulder motion and proper shoulder movement

- Aids the bicep muscle with shoulder biomechanics and reduces the possibility of impingement syndrome

- Directly affects shoulder stability during all active shoulder motions and during throwing activities

- Works to create a stable environment during static shoulder positioning during daily activities

- Aids in 2:1 ratio with shoulder motions

- Imperative in the rehabilitation of the shoulder during the rehabilitation after rotator cuff and shoulder arthroplasty

References

1. Bassett RW, Browne AO. Glenohumeral muscle force and moment mechanics in a position of shoulder instability. *J Biomech.* 1990;23(5):405-415.
2. Szyluk K, Jasinski A. Subacromial impingement syndrome—most frequent reason of the painful shoulder syndrome. *Pol Merkur Lekarski.* 2008;25;146:179-183.
3. Reinold MM, Wilk KE, Fleisig GS. Electromyographic analysis of the rotator cuff and deltoid musculature during common shoulder external rotation exercises. *J Orthop Sports Phys Ther.* 2004;34(7):385-394.

4. Reinold MM, Macrina LC, Wilk KE. Electromyographic analysis of the supraspinatus and deltoid muscles during 3 common rehabilitation exercises. *J Athl Train*. 2007;42(4):464-469.
5. Ryu RK, McCormick J. An electromyographic analysis of shoulder function in tennis players. *Am J Sports Med*. 1988;16(5):481-485.
6. de Oliveira AS, de Morais Carvalho M. Activation of the shoulder and arm muscles during axial load exercises on a stable base of support and on a medicine ball. *J Electromyogr Kinesiol*. 2008;18(3):472-479.
7. Anton D, Shibley LD, Fethke NB. The effect of overhead drilling position on shoulder moment and electromyography. *Ergonomics*. 2001;44(5):489-501.
8. Michiels I, Bodem F. The deltoid muscle: an electromyographical analysis of its activity in arm abduction in various body postures. *Int Orthop*. 1992;16(3):268-271.
9. Rockwood CA, Matsen FA, Wirth MA, Harryman DT. *The Shoulder*. 2nd ed. Philadelphia: W.B. Saunders Company; 1998.
10. Dutton M. *Orthopaedic Examination, Evaluation, and Intervention*. New York: McGraw-Hill; 2008.
11. Clemente, CD. *Gray's Anatomy*. 30th ed. Philadelphia: Lea and Febiger; 1985.
12. Lazarus, MD, Sidles JA. Effect of a chondral-labral defect on glenoid concavity and glenohumeral stability. A cadaveric model. *J Bone Joint Surg Am*. 1996;78;1:94-102.
13. Lippitt S, Matsen F. Mechanisms of glenohumeral joint stability. *Clin Orthop Relat Res*. 1993;291:20-28.
14. Snyder SJ, Karzel RP. SLAP lesions of the shoulder. *Arthroscopy*. 1990;6(4):274-279.
15. Inman VT, Saunders JB. Observations of the function of the shoulder joint. 1944. *Clin Orthop Relat Res*. 1996;330:3-12.
16. Standring S. *Gray's Anatomy: The Anatomical Basis of Clinical Practice*. 40th ed. New York: Churchill Livingstone; 2008.
17. Warfel, J. *The Extremities; Muscles and Motor Points*. Philadelphia: Lea and Febiger; 1985.
18. Buck FM, Jost B. Shoulder arthroplasty. *Eur Radiol*. 2008;18(12):2937-2948.
19. Bhagat S, Bansal M, Sharma H, et al. A rare case of progressive bilateral congenital abduction contracture with shoulder dislocations treated with proximal deltoid release. *Arch Orthop Trauma Surg*. 2008;128(3):293-296.
20. Perry J. Anatomy and biomechanics of the shoulder in throwing, swimming, gymnastics, and tennis. *Clin Sports Med*. 1983;2(2):247-270.
21. Perlmutter GS, Apruzzese W. Axillary nerve injuries in contact sports: recommendations for treatment and rehabilitation. *Sports Med*. 1998;26(5):351-361.
22. Perlmutter GS. Axillary nerve injury. *Clin Orthop Relat Res*. 1999;368:28-36.
23. Berry H, Bril V. Axillary nerve palsy following blunt trauma to the shoulder region: a clinical and electrophysiological review. *J Neurol Neurosurg Psychiatry*. 1982;45(11):1027-1032.
24. Ilaslan H, Iannotti JP. Deltoid muscle and tendon tears in patients with chronic rotator cuff tears. *Skeletal Radiol*. 2007;36(6):503-507.
25. Abrams JS. Arthroscopic approach to massive rotator cuff tears. *Instr Course Lect*. 2006;55:59-66.
26. Nidecker A, Hartweg H. Rare localizations of calcifying tendopathies. *Rofo*. 1983;139(6):658-662.

$$\boxed{3}$$

THE ROTATOR CUFF AND WHAT TRULY CONTROLS THE SHOULDER

The rotator cuff is a muscle system that controls movement of the upper extremity and will help to stabilize the glenohumeral joint. The structure of the muscle will also allow other shoulder muscles, like the deltoid, to work optimally. For example, as you throw a baseball or lift your arm away from your body, the rotator cuff is engaged into action. This occurs along with the scapular stabilizing muscles (eg, the trapezius, rhomboids, and serratus anterior), which will help move your arm in the environment. The rotator cuff is as important to your shoulder as your deep back muscles are for segmental stability in the spine, yet the rotator cuff relies on the coordinated activity of the scapular stabilizers to perform its tasks. When the cuff works in concert with the other shoulder muscles, you will be able to move and function at an optimal level and avoid possible injury.

As you move around in your daily routines and gyrations, the shoulder will go through many quadrants and will require multidirectional stability.[1] To make this happen, the rotator cuff will fire in a synchronistic pattern, producing stability and reducing movement errors that occur during functional activities along with the trapezius, serratus anterior, rhomboids, and the levator scapulae.[2,3] Together they work to form a union that will control shoulder motion in a rhythmical pattern; the rotator cuff's main job is to provide dynamic stability for the glenohumeral joint. This chapter will outline the rotator cuff and its function.

Function of the Rotator Cuff

The rotator cuff consists of 4 muscles, known collectively as the SITS muscles (Figures 3-1, 3-2, 3-3, and 3-4). Their attachments in relation to the humeral head from top to bottom consist of the supraspinatus, infraspinatus, teres minor, and subscapularis. For example, if you are looking at the left shoulder from the posterior direction, the above muscles will be situated in a counterclockwise fashion and then clockwise for the right shoulder. Three of the SITS muscles cause external (or lateral) rotation. Only the subscapularis produces internal rotation of the humerus.[4-6] The function of the rotator cuff as a group is to compress and depress the humeral head within the glenoid cavity during movement[7] by creating an inward and downward pull on the head of the humerus. Stabilizing the humerus dynamically in

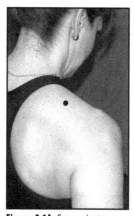

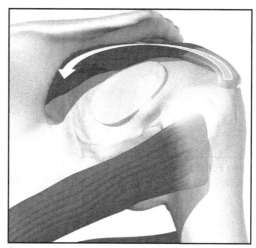

Figure 3-1A. Supraspinatus muscle shown with the muscle belly shown by the black dot.

Figure 3-1B. Supraspinatus schematic with muscle forces shown by the arrow. (Reprinted with permission from Primal Pictures, 2009.)

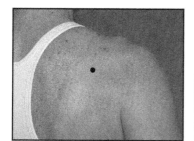

Figure 3-2A. The right infraspinatus with the muscle belly marked by the black dot.

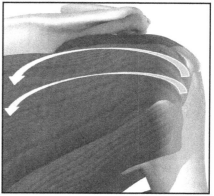

Figure 3-2B. Schematic of the right infraspinatus with muscle forces shown by the arrows. (Reprinted with permission from Primal Pictures, 2009.)

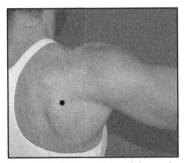

Figure 3-3A. The teres minor and the muscles noted by the black dot.

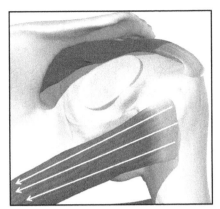

Figure 3-3B. Schematic of the teres minor and the muscle forces from insertion to origin noted by the arrows. (Reprinted with permission from Primal Pictures, 2009.)

Figure 3-4A. The anterior insertion of the right subscapularis at the humeral attachment as noted by the black dot.

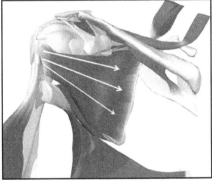

Figure 3-4B. Schematic of the right subscapularis with the muscle forces shown by the arrows from insertion to origin. (Reprinted with permission from Primal Pictures, 2009.)

this way allows other muscles to move the upper extremity without creating impingement within the shoulder joint.[4] **The biomechanical forces of the rotator cuff offset the tendency to produce superior translation of the humeral head as the deltoid abducts the shoulder.** Gravitational forces are not enough to pull the humeral head inward and downward into the glenoid cavity during active motion. Without rotator cuff stabilizing action, shoulder impingement would occur because the humeral head would be forced upward into the coracoacromial ligament, which forms a roof over the supraspinatus tendon and head of the humerus.[7,8]

Origin and Insertion

Origin of the Rotator Cuff (External and Internal Rotators)	Insertion of the Rotator Cuff (External and Internal Rotators)
EXTERNAL ROTATORS	*EXTERNAL ROTATORS*
Supraspinatus: This muscle arises from the supraspinatus fossa on the scapulae (see Figure 3-1).	This muscle will travel across the top of the shoulder to converge deep under the acromion to attach on the superior facet of the greater tuberosity of the humerus (see Figure 3-1).[5,7,9]
Infraspinatus: This triangular muscle originates from the infraspinatus fossa on the scapulae (see Figure 3-2).[5]	The muscle travels from its origin to the middle facet on the greater tuberosity on the humerus (see Figure 3-2).[6,9]
Teres Minor: This is the smallest of the rotator cuff external rotators. The muscle looks like a cylinder and arises from the axillary (lateral) border of the scapulae (see Figure 3-3).	The muscle will travel obliquely upward and laterally to its humeral head attachment on the humeral head at the lower facet and the shoulder joint capsule (see Figure 3-3).[5,6]
Internal Rotator	**Insertion of the Subscapularis**
Origination	
SUBSCAPULARIS The subscapularis is a very large triangular muscle that occupies the entire subscapular fossa (costal surface) of the scapulae (see Figure 3-4).[5,9]	The muscle fibers will course laterally to converge anteriorly on the humeral head and attach to the lesser tuberosity, intertubercular groove, and the anterior portion of the shoulder capsule (see Figure 3-4).[5,6,9]
Action of the External and Internal Rotators	
EXTERNAL ROTATORS The 3 SITS muscles cause external (or lateral) rotation of the humeral head.[4-6]	
INTERNAL ROTATORS The subscapularis produces internal rotation of the humerus.[4-6]	

Pathomechanics of Common Injuries

A player comes off the bench, gets checked in to the boards, and then comes into the locker room with so-called "shoulder pain." **Shoulder pain** is common among athletes[10,11] and also among the "normal" population from a variety of situations (ie, overuse, too much weekend sports, or even just sleeping funny).

The rotator cuff is often injured in the shoulder and the **supraspinatus is the most common of the tendons that are involved.**[12-14] Namdari et al[12] showed that the supraspinatus tendon is compromised or torn most commonly in its anterosuperior aspect. Other tears also included the infraspinatus and sometimes the long head of the biceps brachii, depending upon the severity of the injury and the mechanism. Clearly, poorer outcomes are associated with multiple trauma(s) to a multitude of tendons and tissues versus tears that involve the supraspinatus only.

Further injuries that involve the shoulder and the rotator cuff muscles can include other structures that surround the humeral head (ie, fractures).[15] When a person falls on his or her arm and hits the tip of his or her shoulder, **fracturing the greater tubercle**, in many cases this will cause a tear in the rotator cuff. Even daily life activities can cause problems, and with the constant motion of the shoulder in a typical day, it is not surprising that muscle tears, whether it is the supra or infraspinatus, can change the mechanics of motion and function.[16] Indeed, it is easy to understand that the supraspinatus is the most commonly torn muscle/tendon system around the shoulder, usually due to subacromial impingement.[11,17,18] After the initial injury, further trauma often occurs when resulting instability causes **impingement** of the subscapularis against the coracoid process.[19] When that happens, patients can experience further tearing, not only in the supraspinatus, but possibly involving the infraspinatus and/or subscapularis. This type of multitrauma will lead to significant pain and poor shoulder function.[19]

Clinical Pearls: Rotator Cuff Muscles

- The SITS muscles are very significant for normal shoulder biomechanics during daily life
- Depresses the humeral head during flexion and abducted motions and moves the humeral head in an anterior direction during external rotation

- The rotator cuff along with the long head of the biceps will work to keep the humeral head in the glenoid and prevent it from moving too far superiorly during motion

- External rotators will provide lateral rotational power and will pull the humeral head into the glenoid for support and proper motion

- The internal rotators will help with internal rotation and will aid with humeral head posterior translation during motion

References

1. Carpenter MG, Frank JS. Influence of postural anxiety on postural reactions to multidirectional surface rotations. *J Neurophysiol.* 2004;92(6):3255-3265.

2. de Groot JH, Rozendaal LA. Isometric shoulder muscle activation patterns for 3-D planar forces: a methodology for musculo-skeletal model validation. *Clin Biomech (Bristol, Avon).* 2004;19(8):790-800.

3. Lin JC, Weintraub N. Nonsurgical treatment for rotator cuff injury in the elderly. *J Am Med Dir Assoc.* 2008;9(9):626-632.

4. Millett PJ, Wilcox RB 3rd. Rehabilitation of the rotator cuff: an evaluation-based approach. *J Am Acad Orthop Surg.* 2006;14(11):599-609.

5. Clemente CD. *Gray's Anatomy: The Anatomical Basis of Clinical Practice.* 30th ed. Philadelphia: Lea and Febiger; 1985.

6. Standring S. *Gray's Anatomy: The Anatomical Basis of Clinical Practice.* 40th ed. New York: Churchill Livingstone; 2008.

7. Dutton M. *Orthopaedic Examination, Evaluation, and Intervention.* New York: McGraw-Hill; 2008.

8. Alon G. Shoulder Kinematic and Pathological Movement Patterns; Course Notes. University of Maryland; 2003.

9. Warfel J. *The Extremities: Muscles and Motor Points.* Philadelphia: Lea and Febiger; 1985.

10. Cobiella CE. Shoulder pain in sports. *Hosp Med.* 2004;65(11):652-656.

11. Jobe FW, Pink M. The athlete's shoulder. *J Hand Ther.* 1994;7(2):107-110.

12. Namdari S, Henn RF 3rd. Traumatic anterosuperior rotator cuff tears: the outcome of open surgical repair. *J Bone Joint Surg Am.* 2008;90(9):1906-1913.

13. Ecklund KJ, Lee TQ. Rotator cuff tear arthropathy. *J Am Acad Orthop Surg.* 2007;15(6):40-349.

14. Hata Y, Saitoh S. Volume changes of supraspinatus and infraspinatus muscles after supraspinatus tendon repair: a magnetic resonance imaging study. *J Shoulder Elbow Surg.* 2005;14(6):631-635.

15. Kendall CB, Tanner SL. SLAP tear associated with a minimally displaced proximal humerus fracture. *Arthroscopy.* 2007;23(12):1362.

16. Adams CR, Baldwin MA. Effects of rotator cuff tears on muscle moment arms: a computational study. *J Biomech.* 2007;40(15):3373-3380.

17. Crues JV 3rd, Fareed DO. Magnetic resonance imaging of shoulder impingement. *Top Magn Reson Imaging.* 1991;3(4):39-49.

18. Altchek DW, Dines DM. Shoulder injuries in the throwing athlete. *J Am Acad Orthop Surg.* 1995;3(3):159-165.

19. MacMahon PJ, Taylor DH. Contribution of full-thickness supraspinatus tendon tears to acquired subcoracoid impingement. *Clin Radiol.* 2007;62(6):556-563.

$$\boxed{4}$$

THE TRICEPS AND ITS ROLE IN SHOULDER STABILITY

When you push up from a chair, get up from the floor, or push to get out of your car, you are engaging your triceps brachii. The triceps brachii may be the forgotten arm muscle group since most people think first of the large biceps and shoulder muscles. However, the triceps accounts for valuable function at the shoulder and the elbow and does have an important role with stability.[1,2] In this section, we will outline the triceps function and its role at the shoulder.

Function of the Triceps Brachii

The primary role of the triceps brachii is to extend the elbow and the shoulder. It is composed of 3 heads: the long head, lateral head, and medial head (Figure 4-1).[3-5]

In general terms, all 3 heads of triceps are recruited for a given task and the motor units in each head may be called upon preferentially to provide specific assistance. If a person extends his or her shoulders, the long head is recruited, and when a heavy load is lifted, the long and medial heads are principally used.[3,6,7]

During activities that involve extension of the elbow, most of the power comes from the long head and the medial head, with the lateral head acting as a support.[3,8] When a person is performing an activity such as hiking or pulling him- or herself upward, the triceps will be activated.[9] During sporting activities that require throwing or even gymnastics, there is a strong triceps requirement that must occur for proper arm functioning.[10]

As with most muscles, the triceps can be activated more strongly voluntarily, under control of the nervous system, than by electrical stimulation.[11] If a person is lacking in triceps strength from an injury, electrical stimulation will help to recruit the muscle fibers, but the stimulation will aid the internal workings of the nervous system for proper muscle firing.[12] From a clinical perspective, we would like to encourage a person to recruit his or her muscles during exercise and use stimulation only as an adjunct to voluntary effort. This chapter will discuss triceps function, origin and insertion, and pathomechanics.

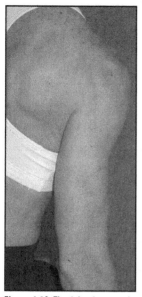

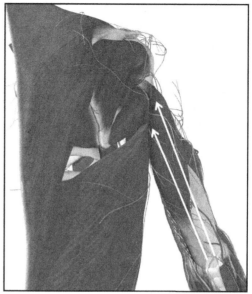

Figure 4-1A. The right tricep muscle in sitting.

Figure 4-1B. Schematic with the muscle forces noted by the arrows from insertion at the elbow to origin in the shoulder. (Reprinted with permission from Primal Pictures, 2009.)

Origin and Insertion

Origin and Insertion for the Tricep Muscle	Insertion of the Tricep Muscle
ORIGIN	*INSERTION*
THE LONG HEAD OF THE TRICEP	
The long head of the triceps brachii arises from a flat tendon at the infraglenoid tubercle of the scapulae (see Figure 4-1).	The long head has its muscle fibers travel between the medial and lateral heads to join them as a conjoint tendon to attach to the olecranon process of the ulna at the elbow (see Figure 4-1).[4,5,7,13]
THE LATERAL HEAD OF THE TRICEP	
The lateral head of the triceps brachii arises from the posterior and lateral surface of the humerus (see Figure 4-1).	Its fibers travel downward and attach with the medial and long head as a conjoint tendon on the ulna (see Figure 4-1).

(Continued)

THE MEDIAL HEAD OF THE TRICEP The medial head of the triceps brachii arises from the medial side of the humerus and the medial intermuscular septum[6,7] and is considered a deep muscle (see Figure 4-1).	Its fibers descend and attach with the conjoint tendon to the posterior aspect of the olecranon of the ulna (see Figure 4-1).

ACTION OF THE TRICEP MUSCLE

The primary function of the triceps is to extend the elbow. Since the long head attaches to the scapulae and crosses the posterior aspect of the shoulder joint, it will also aid in extending and adducting the humerus at the glenohumeral joint.[3,5,7,14]

Pathomechanics of Common Injuries

According to Dutton,[3] and supported by other researchers,[15] **the most common elbow injury is triceps tendinitis.**[3,16] The result of this overuse injury is irritation of the triceps tendon with subsequent swelling sometimes noted.[3,14,15] Another injury that can occur around the elbow and the tricep muscle is **olecranon bursitis.**[17] The trouble with this injury is that once the bursa is inflamed, it can become septic[18] and this can lead to poor range of motion and soreness with daily activities. The triceps can also become subject to eccentric loading injuries during controlled flexion (lowering yourself downward from an elbow semi-bent position, or during a bench press) if the load is too great.[19,20]

A rare injury to the tricep muscle is a **partial- or full-thickness tear** of the triceps, and occasionally, an avulsion, or tearing of the triceps tendon from its insertion at the olecranon process.[12,21-23]

Clinical Pearls: Tricep Muscle

- Supports the posterior shoulder joint and will aid in pushing up from a chair, wheelchair, or from the floor

- The tricep muscle will be active during the release portion of the throwing motion of sporting activities

- The tricep will help support the upper arm when the distal segment is fixed (ie, gymnastics floor routines)

References

1. Bassett RW, Browne AO. Glenohumeral muscle force and moment mechanics in a position of shoulder instability. *J Biomech.* 1990;23(5):405-415.
2. Rockwood CA, Matsen FA, Wirth MA, Harryman DT. *The Shoulder.* 2nd ed. Philadelphia: W.B. Saunders Company; 1998.
3. Dutton M. *Orthopaedic Examination, Evaluation, and Intervention.* New York: McGraw-Hill; 2008.
4. Madsen M, Marx RG, Millett PJ. Surgical anatomy of the triceps brachii tendon: anatomical study and clinical correlation. *Am J Sports Med.* 2006;34(11):1839-1843.
5. Warfel J. *The Extremities: Muscles and Motor Points.* Philadelphia: Lea and Febiger; 1985.
6. Belentani C, Pastore D, Wangwinyuvirat M, et al. Triceps brachii tendon: anatomic-MR imaging study in cadavers with histologic correlation. *Skeletal Radiol.* 2009;38(2):171-175.
7. Clemente CD. *Gray's Anatomy.* 30th ed. Philadelphia: Lea and Febiger; 1985.
8. Norkin CC, Levangie PK. *Joint Structure and Function: A Comprehensive Analysis.* 2nd ed. Philadelphia: F.A. Davis Company; 1992.
9. Foissac MJ, Berthollet R. Effects of hiking pole inertia on energy and muscular costs during uphill walking. *Med Sci Sports Exerc.* 2008;40(6):1117-1125.
10. Terzis G, Karampatsos G. Neuromuscular control and performance in shot-put athletes. *J Sports Med Phys Fitness.* 2007;47(3):284-290.
11. Huffenus AF, Forestier N. Effects of fatigue of elbow extensor muscles voluntarily induced and induced by electromyostimulation on multi-joint movement organization. *Neurosci Lett.* 2006;403(1-2):109-113.
12. Langenhan R, Weihe R. Traumatic rupture of the triceps brachii tendon and ipsilateral Achilles tendon. *Unfallchirurg.* 2007;110(11):977-980.
13. Standring S. *Gray's Anatomy: The Anatomical Basis of Clinical Practice.* 40th ed. New York: Churchill Livingstone; 2008.
14. Guerroudj M, de Longueville JC. Biomechanical properties of triceps brachii tendon after in vitro simulation of different posterior surgical approaches. *J Shoulder Elbow Surg.* 2007;16(6):849-853.
15. Ervilha UF, Arendt-Nielsen L. The effect of muscle pain on elbow flexion and coactivation tasks. *Exp Brain Res.* 2004;156;2:174-182.
16. Fulcher SM, Kiefhaber TR. Upper-extremity tendinitis and overuse syndromes in the athlete. *Clin Sports Med.* 1998;17(3):433-448.
17. Tran N, Chow K. Ultrasonography of the elbow. *Semin Musculoskelet Radiol.* 2007;11(2):105-116.
18. Laupland KB, Davies HD. Olecranon septic bursitis managed in an ambulatory setting. The Calgary Home Parenteral Therapy Program Study Group. *Clin Invest Med.* 2001;24(4):171-178.
19. Holleb PD, BR Bach BR Jr. Triceps brachii injuries. *Sports Med.* 1990;10(4):273-276.
20. Cooney WP 3rd. Sports injuries to the upper extremity. How to recognize and deal with some common problems. *Postgrad Med.* 1984;76(4):45-50.
21. Sonin A. Tendon disorders. *Semin Musculoskelet Radiol.* 1998;2(2):163-174.
22. Pina A, Garcia I. Traumatic avulsion of the triceps brachii. *J Orthop Trauma.* 2002;16(4):273-276.
23. Wagner JR, Cooney WP. Rupture of the triceps muscle at the musculotendinous function: a case report. *J Hand Surg Am.* 1997;22;2:341-343.

5

THE BICEPS AND ITS ROLE IN THE SHOULDER

The biceps brachii muscle is a large functional muscle with 2 heads that originate from the scapula (Figure 5-1). The bicep muscle is much more than just a flexor of the elbow. Biceps also contributes, in concert with the rotator cuff and scapular stabilizers, to move the humeral head safely and securely. This chapter will deal with the biceps and how it affects a person's arm whether in movement or stability.

Function of the Biceps

The biceps contributes forces to several joints, including the elbow, the radioulnar joints, and the glenohumeral joint. The bicep will contribute to flexion, supination at the elbow, and flexion at the shoulder.[1-5] The biceps' effect on the shoulder is less well understood than its effect at the elbow.[1] Through its long head attachment, the biceps can depress the head of the humerus along with the rotator cuff, thus improving anterior shoulder stability.[1,3,6]

The biceps' functions at the shoulder include tensing the glenoid labrum during elevation to avoid pinching, limiting external rotation, and compressing the humeral head into the glenoid cavity.[7-10] The long head of the biceps thus provides stability during abduction and rotation (up to 90 degrees) of the humerus. The long head will help to pull the humeral head into the glenoid cavity and is an important, if underrated, stabilizer of the shoulder joint during motion.[1] As long as its attachments are intact, the biceps acts synergistically with the rotator cuff muscles to maintain the health of the glenohumeral joint by adding a depressive and compressive force to the humeral head.[4,11,12]

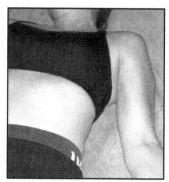

Figure 5-1A. Relaxed bicep in vivo in supine.

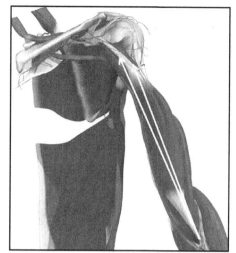

Figure 5-1B. Schematic of the bicep muscle with arrow shows the muscle forces from insertion to origin. (Reprinted with permission from Primal Pictures, 2009.)

Origin and Insertion

Origin of the Biceps Brachii	Insertion of the Biceps Brachii
The Long Head of the Biceps	
The long head arises from the supraglenoid tubercle of the scapulae, the glenoid labrum, and the shoulder capsule (see Figure 5-1).[1,5,13,14]	The long head travels downward to join the short head of the biceps and both attach distally on the bicipital tuberosity of the radius (Figure 5-2).
The Short Head of the Biceps	
The short head of biceps originates from the coracoid process via a flattened tendon it shares with 2 other muscles, coracobrachialis and pectoralis minor (see Figure 5-1).	The short head of the biceps will course downward, joining with the long head to attach distally on the bicipital tuberosity (see Figure 5-2).[1,5,14]
The Action of the Biceps Brachii	
The elbow and radioulnar components of the bicep muscle group will give rise to elbow flexion, supination, and limited pronation.[1,4,15]	

Pathomechanics of Common Injuries

Common bicep injuries that clinicians commonly see in the clinic:

- Bicep tendinitis[16]
- Bicep calcified tendinitis[17,18]

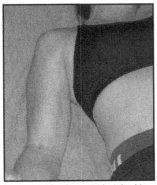

Figure 5-2A. Contracted right bicep muscle in vivo in supine.

Figure 5-2B. Schematic of the bicep long head with muscle forces shown by the arrow from insertion to origin. (Reprinted with permission from Primal Pictures, 2009.)

- Bicipital impingement or subluxation[18-20]

- Labral involvement[21-23]

- Rotator cuff dysfunction[24]

Bicep tendinitis is caused by, in most cases, overuse of a person's arm with his or her forearm either in pronation or supination.[1] The overuse can involve weightlifting, pulling activities (like waterskiing), or sports that cause a person to forcibly use his or her arm.[25]

Bicep calcified tendinitis is a condition that occurs most in females[26,27] and usually occurs when a person is in his or her 40s and 50s. The bicep has calcium hydroxyapatite crystal deposited in it and is stated to be a result of trauma or overuse.[27]

Bicipital impingement or subluxation will involve the long head of the bicep tendon and it moves from its position in the bicipital groove to a medial position.[28] Repetitive movement of the biceps from the center of the intratubercular groove to the medial side will usually be accompanied by formation of a bony spur in the groove, which can cause the long head to become frayed, requiring surgical reattachment.[28,29] In many cases, biceps pathology will accompany a rotator cuff tear. When they occur together, they cause the shoulder to become very sore and unstable.[16]

Labral involvement as a result of injury to the long head of the biceps is due to the location and orientation of the long head and its attachment to the superior anchor of the labrum.[19,21,30-32] The superior labral attachment is affected when the long head of the bicep is overstrained.[32,33] Overstrain can occur during a sporting activity[34] or when a person has a pre-existing co-morbidity (health condition) that would affect the stability of the shoulder (ie, stroke, rheumatoid arthritis).[35] The superior labrum can also be injured when a person falls on his or her outstretched hand and arm or with a pulling injury.[16]

Rotator cuff dysfunction with biceps pathology is common[36] and it is the most commonly occurring injury of any tendon in the human body.[37] Rotator cuff tears can have a variety of tendons involved (remember the SITS [supraspinatus, infraspinatus, teres minor, and subscapularis] muscles) and the severity of the injury will dictate how many of the tendons are affected.[19,38-41]

Clinical Pearls: Bicep

- Biceps brachii will significantly contribute to humeral head depression along with the rotator cuff muscles

- Will provide support anteriorly for the shoulder and help to keep the humeral head from moving to far forward during daily activities and also during sports (ie, throwing a ball)

- Aids in biomechanics with the rotator cuff to give the humeral head proper arthrokinematics

- Reduces wear on the humeral head with its function by not letting anterior translation take place in an excessive manner and reinforces the function of the rotator cuff from an anterior position

- Elbow flexors but the strongest portion of the muscle mass is from 80 to 100 degrees

- Supinator of the forearm and will aid in daily life functions that deal with forearm motions (ie, opening a door, turning a key, driving)

- Acts as a semipronator of the forearm and again will aid in daily life activities of typing, pulling on pants, hygiene, and during sporting activities

References

1. Dutton M. *Orthopaedic Examination, Evaluation, and Intervention*. New York: McGraw-Hill; 2008.

2. Bazzucchi I, Sbriccoli P, et al. Coactivation of the elbow antagonist muscles is not affected by the speed of movement in isokinetic exercise. *Muscle Nerve.* 2006;33(2):191-199.

3. Norkin CC, Levangie PK. *Joint Structure and Function: A Comprehensive Anaylsis.* 2nd ed. Philadelphia: FA Davis Co; 1992.

4. Warner JJ, McMahon PJ. The role of the long head of the biceps brachii in superior stability of the glenohumeral joint. *J Bone Joint Surg Am.* 1995;77(3):366-372.

5. Clemente CD. *Gray's Anatomy.* 30th ed. Philadelphia: Lea and Febiger; 1985.

6. Neer CS. Second anterior acromioplasty for the chronic impingement syndrome in the shoulder: a preliminary report. *J Bone Joint Surg Am.* 1972;54(1):41-50.

7. Hurley JA, Anderson TE. Shoulder arthroscopy: its role in evaluating shoulder disorders in the athlete. *Am J Sports Med.* 1990;18(5):480-483.

8. Basmajian JV, Bazant FJ. Factors preventing downward dislocation of the adducted shoulder joint. An electromyographic and morphological study. *J Bone Joint Surg Am.* 1959;41-A:1182-1186.

9. Basmajian JV, Deluca CJ. *Muscles Alive: Their Functions Revealed by Electromyography.* Baltimore, MD: Williams & Wilkins; 1985.

10. Itoi E, Kuechle DK, et al. Stabilising function of the biceps in stable and unstable shoulders. *J Bone Joint Surg Br.* 1993;75(4):546-550.

11. Payne LZ, Altchek DW, et al. Arthroscopic treatment of partial rotator cuff tears in young athletes. A preliminary report. *Am J Sports Med.* 1997;25(3):299-305.

12. Kido T, Itoi E, et al. The depressor function of biceps on the head of the humerus in shoulders with tears of the rotator cuff. *J Bone Joint Surg Br.* 2000;82(3):416-419.

13. Standring S. *Gray's Anatomy: The Anatomical Basis of Clinical Practice.* 40th ed. New York: Churchill Livingstone; 2008.

14. Warfel J. *The Extremities: Muscles and Motor Points.* Philadelphia: Lea and Febiger; 1985.

15. Rockwood CA, Matsen FA. *The Shoulder.* 2nd ed. Philadelphia: WB Saunders Co; 1998.

16. DeLee JC, Drez D Jr. *Orthopaedic Sports Medicine: Principles and Practice.* Vol 2. Philadelphia: WB Saunders Co; 1994.

17. Cone RO 3rd, Resnick D, et al. Shoulder impingement syndrome: radiographic evaluation. *Radiology.* 1984;150(1):29-33.

18. Cone RO, Danzig L, et al. The bicipital groove: radiographic, anatomic, and pathologic study. *Am J Roentgenol.* 1983;141(4):781-788.

19. Stubbs SN, Hunter RE. Complete, superior labral radial tear and type II slap tear associated with greater tuberosity fracture. *Arthroscopy.* 2004;20(Suppl 2):70-72.

20. MacMahon PJ, Taylor DH, et al. Contribution of full-thickness supraspinatus tendon tears to acquired subcoracoid impingement. *Clin Radiol.* 2007;62(6):556-563.

21. Altchek DW, Dines DM. Shoulder injuries in the throwing athlete. *J Am Acad Orthop Surg.* 1995;3(3):159-165.

22. Araghi A, Prasarn M, et al. Recurrent anterior glenohumeral instability with onset after forty years of age: the role of the anterior mechanism. *Bull Hosp Jt Dis.* 2005;62(3-4):99-101.

23. Tomlinson RJ Jr, Glousman RE. Arthroscopic debridement of glenoid labral tears in athletes. *Arthroscopy.* 1995;11(1):42-51.

24. Sakurai G, Ozaki J, et al. Morphologic changes in long head of biceps brachii in rotator cuff dysfunction. *J Orthop Sci.* 1998;3(3):137-142.

25. Morrey MF. *The Elbow and Its Disorders.* 2nd ed. Philadelphia: WB Saunders Co; 1993.

26. Uhthoff HK, Sarkar K. Calcifying tendinitis. *Baillieres Clin Rheumatol.* 1989;3(3):567-581.

27. Uhthoff HK, Sarkar K. An algorithm for shoulder pain caused by soft-tissue disorders. *Clin Orthop Relat Res.* 1990;(254):121-127.

28. Neviaser RJ. Lesions of the biceps and tendinitis of the shoulder. *Orthop Clin North Am.* 1980;11(2):343-348.

29. Farin PU, Jaroma H, et al. Shoulder impingement syndrome: sonographic evaluation. *Radiology.* 1990;176(3):845-849.

30. Jobe FW, Pink M. The athlete's shoulder. *J Hand Ther.* 1994;7(2):107-110.

31. Rafii M, Firooznia H, et al. CT arthrography of capsular structures of the shoulder. *Am J Roentgenol.* 1986;146(2):361-367.

32. Burkhart SS, Morgan CD, et al. Shoulder injuries in overhead athletes. The "dead arm" revisited. *Clin Sports Med.* 2000;19(1):125-158.

33. Park JH, Lee YS, et al. Outcome of the isolated SLAP lesions and analysis of the results according to the injury mechanisms. *Knee Surg Sports Traumatol Arthrosc.* 2008;16(5):511-515.

34. Hurley JA, Anderson TE. Shoulder arthroscopy: its role in evaluating shoulder disorders in the athlete. *Am J Sports Med.* 1990;18(5):480-483.

35. Shah RR, Haghpanah S, et al. MRI findings in the painful poststroke shoulder. *Stroke.* 2008;39(6):1808-1813.

36. Ecklund KJ, Lee TQ, et al. Rotator cuff tear arthropathy. *J Am Acad Orthop Surg.* 2007;15(6):340-349.

37. Lohr JF, Uhthoff HK. Epidemiology and pathophysiology of rotator cuff tears. *Orthopade.* 2007;36(9):788-795.

38. Steenbrink F, de Groot JH, et al. Pathological muscle activation patterns in patients with massive rotator cuff tears, with and without subacromial anaesthetics. *Man Ther.* 2006;11(3):231-237.

39. Olive RJ Jr, Marsh HO. Ultrasonography of rotator cuff tears. *Clin Orthop Relat Res.* 1992;(282):110-113.

40. Lee SB, Nakajima T, et al. The bursal and articular sides of the supraspinatus tendon have a different compressive stiffness. *Clin Biomech (Bristol, Avon).* 2000;15(4):241-247.

41. Luo ZP, Hsu HC, et al. Mechanical environment associated with rotator cuff tears. *J Shoulder Elbow Surg.* 1998;7(6):616-620.

$$\boxed{6}$$

THE FOREARM AND HAND MUSCLES: WHAT IS THEIR ROLE?

Most of us have the ability to get up every day and move around using our legs, but to really appreciate the wonders of the body, look at what the hand and forearm can do. The hand can move in a multitude of directions and can also perform very fine motor activities with amazing precision. Your hand gives you the ability to play the violin, wash your car, count money, and throw a baseball. All of these activities take coordination and dexterity that come from an incredible orchestration between your central nervous system and your forearm and hand musculature.

When you play the violin, guitar, or the piano, you are using an array of muscles that make the task look easy but the incredible part is that all of the various motor units are turned on and off and on again in milliseconds, almost without conscious intervention. We are uniquely made, all of us, but we are alike in the sense that our muscles are controlled by a nervous system, and that nervous system relies on feedback from the muscles, ligaments, and joints to provide a coordinated response to life's challenges. As we go through life, swimming, throwing a ball, shooting a hockey puck, riding a bike, or pushing a wheelchair, a coordinated effort occurs in all our muscles to allow us to remain upright and move our arms and hands. This chapter is devoted to the forearm and intrinsic hand muscles.

Function of the Forearm and Hand Muscles

The hand is a part of the body that has been estimated to account for as much as 90% of a person's function.[1] The thumb is responsible for 40% to 50% of hand function[1] and together the hand and wrist are used in a variety of ways during daily activities. The hand and wrist are intimately related both anatomically and functionally. Proximally, the wrist is the joint between the distal radius and radioulnar disc and 8 carpal bones. The bony hand consists of the carpals, their articulations with the metacarpals, the long bones that make up the palm, and the phalanges (shorter bones that make up the fingers). The large muscles reaching the fingers cross over the distal radioulnar joint (DRUJ) and the wrist in their course. Dysfunction at either joint can thus affect hand function. The DRUJ is important in producing one of the most functional motion patterns of the forearm—pronation/supination—which is related to motions of the upper extremity used for eating, bathing, and a multitude of other tasks.[1,2] If the DRUJ has limited motion, not only will pronation be diminished, but wrist flexion and extension will also not be optimal.[2] The DRUJ is bound together distally at the radioulnar disc, sometimes known as the triangular fibrocartilage complex (TFCC). The TFCC helps to stabilize the radius to the ulna and also limits supination and pronation.[2] Moreover, the TFCC serves as the ulnar portion of the proximal wrist joint and articulates with the carpal bones, intervening between them and the distal ulna, which is largely nonarticulatory.

The carpals are arranged in 2 rows from proximal to distal. The distal row consists of the hamate, capitate, trapezoid, and trapezium (from ulnar side to the radial) and the proximal row consists of triquetrum with its anteriorly placed pisiform, lunate, and scaphoid, again going from ulnar to radial.[1-3] The pisiform serves as an attachment for the tendon of the flexor carpi ulnaris muscle and is thus a sesamoid bone. It is also one of the 4 anchors for the carpal tunnel at the wrist, along with the hook of the hamate, the tubercle of the scaphoid, and the trapezium. The pisiform can be easily palpated as a mobile mass on the ulnar side of the hand by slightly flexing the relaxed wrist.

The wrist is responsible for motions including flexion, extension, radial and ulnar deviation, and circumduction, a circular movement that is a combination of all the others.[1-4] To make the wrist perform the activities listed above requires neuromuscular control of 24 extrinsic and 19 intrinsic muscles associated with the hand and forearm, not to mention coordinated activity of elbow, shoulder, and scapular muscles (Figures 6-1, 6-2, 6-3, and 6-4).[1-3] To cause the wrist to move into extension, muscles on the posterior side of the forearm must be engaged (see Figures 6-1 and 6-2).[3,5] To have the wrist perform flexion, muscles on the anterior (flexor) side of the forearm must contract (see Figure 6-3).[5] The extrinsic hand muscles originate in the forearm, while the intrinsic hand muscles originate within the hand itself. The extrinsic hand muscles are larger and more powerful than their intrinsic counterparts. Flexor muscles and pronator teres originate from the medial side of the humerus at the common flexor tendon. There are muscles specialized for ulnar and radial deviation located both anteriorly (flexor carpi ulnaris and radialis) and posteriorly (extensor carpi ulnaris and radialis) (see Figure 6-4). The wrist can serve as a modulator of muscle activity for the flexors of the fingers. For example, by extending the wrist, the optimal length-tension relationship can be maintained as the flexor muscles shorten over the fingers. Thus our ability to grip an object tightly depends, to some degree, on our ability to extend the wrist.[6]

The elbow, on the other hand, in deference to its main functions of flexion/extension and supination/pronation, does not possess as many degrees

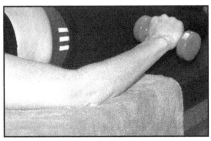

Figure 6-1A. Wrist extensors contracted with a weight in the hand. As the hand is moved into wrist extension, the muscles will contract, causing shortening from the wrist insertion to the elbow.

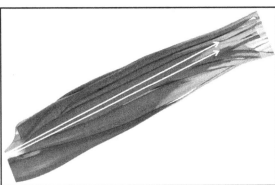

Figure 6-1B. Schematic of forearm extensors with arrows showing the muscle forces from insertion to origin back up to the elbow. (Reprinted with permission from Primal Pictures, 2009.)

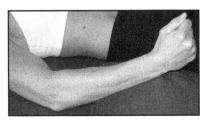

Figure 6-2A. The brachioradialis in vivo contracted with the wrist moving into radial deviation to contract the muscle.

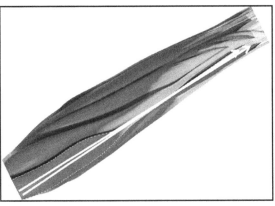

Figure 6-2B. Schematic internal illustration of the forearm brachioradialis muscle with arrow showing the muscle forces from insertion to origin. (Reprinted with permission from Primal Pictures, 2009.)

Figure 6-3A. The forearm flexors in vivo contracted with the wrist into the flexed position with the weight in hand.

Figure 6-3B. Schematic internal illustration of the forearm flexor muscles with muscle forces shown from the insertion to the origin noted by the arrow. (Reprinted with permission from Primal Pictures, 2009.)

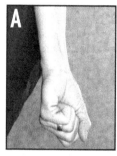

Figure 6-4. (A) The right hand with the thumb and wrist shown in supine. (B) The right abductor pollicis brevis. (C) The right abductor pollicis longus. (D) The extensor pollicis brevis. (E) The extensor pollicis longus. The arrows show the muscle forces for the hand muscles. (Reprinted with permission from Primal Pictures, 2009.)

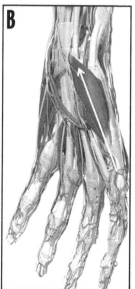

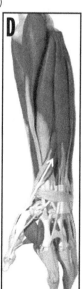

of freedom as the wrist and hand complex.[1,7] The flexion axis is generally agreed to lie through the capitulum and the trochlea on the humerus and is slightly tilted from lateral to medial.[1] Motion into flexion can go as high as 145 degrees limited by the bulk of the elbow flexor muscles, which results in a tissue contact endfeel. Supination and pronation take place at the proximal and distal radioulnar joints, with a nontrivial contribution from lateral and medial gliding of the ulna at the elbow joint.[1,8] The longitudinal axis of motion for pronation and supination lies on a line through the radial head to the center of the inferior radial ulnar joint.[9,10] The muscles that control the forearm and subsequently the wrist and hand, are the forearm flexors and extensors.[1,3,11]

Origin and Insertion

Origin of the Forearm Extensors	Insertion of the Forearm Extensors
Extensor Carpi Radialis Longus (ECRL) The ECRL is beneath the brachioradialis at its origin, arising from the distal third of the supracondylar ridge on the humerus.[1,3,12] The ECRL also attaches to the intermuscular septum on the lateral side and to the common extensor tendon, but only a few fibers originate from there (see Figure 6-1).[3]	The muscle fibers for the ECRL travel distally, becoming tendinous in the middle third of the forearm[3,11] and continue to cross the wrist joint and insert on the radial side of the second metacarpal bone near its base (see Figure 6-1).
Extensor Carpi Radialis Brevis (ECRB) This is a short and thick muscle[3] that will be covered by the ECRL. The ECRB will arise from the lateral epicondyle of the humerus, forming part of the common tendon of the forearm extensors.[3,11,13] The muscle will also arise from the radial collateral ligament and from the intermuscular septum (see Figure 6-1).[3]	The muscle traverses distally and in the middle of the forearm becomes tendinous.[3] The tendon continues downward to cross the wrist joint and attach on the radial side of the base of the third metacarpal (see Figure 6-1).[3,11,13]
Brachioradialis The brachioradialis is an elbow flexor and forms the bulk of the superficial muscles on the extensor side of the forearm.[3,11] The muscle arises from the proximal two-thirds of the lateral supracondylar ridge, between the triceps and brachialis muscles (see Figure 6-2).[1,3,11]	The brachioradialis parallels the radius where it provides cover for the superficial branch of the radial nerve. The muscle fibers change to tendon in the middle of the forearm, and the tendon attaches to the radial styloid process on the lateral side (see Figure 6-2).[1,3,13]

(Continued)

Extensor Digitorum

This is a unique muscle because it is the main extensor for all the digits in the hand and is the extensor counterpart to the muscles flexor digitorum profundus and superficialis.[3] The extensor digitorum arises from the common extensor tendon at the humerus[11] and from the intermuscular septum (see Figure 6-1).[3]

The muscle fibers traverse distally, dividing into 4 tendons that converge on the back of the hand where the tendons attach to the base of the proximal and distal phalanx of each of the 4 fingers.[3,13] The index finger's extensor digitorum tendon will accompany the extensor indicis to attach to the proximal phalanx on the dorsum of the hand.[3,11,13] The attachments of the extensor digitorum on the proximal and middle phalanx mirror the attachments of flexor digitorum superficialis (proximal phalanx) and flexor digitorum profundus (distal phalanx), except that the extensor attachment to the proximal phalanx is a single slip and the distal attachment is a double slip, just the reverse for the flexors (see Figure 6-1).

Extensor Indicis

This muscle arises from the posterior surface of the ulna and the interosseus membrane.

The muscle travels obliquely downward across the hand to attach to the dorsal aspect of the proximal phalanx of the second digit.[3,13]

Extensor Carpi Ulnaris (ECU)

This muscle arises from the lateral epicondyle on the humerus via the common extensor tendon and will also arise for the intermuscular septum and the posterior border of the ulna (see Figure 6-1).[14]

The muscular tendon will pass through the extensor retinaculum and will attach to the base of the medial side of the fifth metacarpal bone.

Action of the Forearm Muscle Group

The action of the forearm extensors is to perform wrist extension and also radial deviation. The ERCL has more of a component to perform radial deviation than the ECRB, but both are involved with both actions. The ECU will also cause wrist extension, but will have an action component of ulnar deviation and even some flexion.[5,14]

Pronator Teres

The muscle arises from 2 origins: the humeral head (more superficial) arises from the humerus on the medial supracondylar ridge and the medial epicondyle via the common flexor tendon. The ulnar head is a thin layer of muscle that originates from the coronoid process of the ulna (see Figure 6-3).[1,3]

Both muscular heads will travel obliquely from their attachment sites to the anterolateral surface of the radius (see Figure 6-3).[1,3,13]

Origin of the Forearm Flexors	Insertion of the Forearm Flexors
Flexor Carpi Radialis (FCR) The FCR arises from the medial epicondyle of the humerus via the common flexor tendon.[1-3] The muscle lies on the ulnar side of the pronator teres muscle (see Figure 6-3).	The FCR muscle fibers travel on the medial side of the forearm, becoming tendinous in the middle of the forearm. The tendon travels down the forearm to the wrist and after attaching to the trapezium in the proximal carpal row, continues to reach the bases of the second and third metacarpal bones.

(Continued)

Origin of the Forearm Flexors	Insertion of the Forearm Flexors
Flexor Carpi Ulnaris (FCU) The FCU muscle lies on the ulnar side of the forearm.[3] This muscle has 2 heads of origin. One of the muscle attachments arises from the medial epicondyle of the humerus.[3,11] The other muscle attachment arises from the olecranon of the ulna.	Both heads converge and travel to attach on the pisiform carpal bone. From there the piso-hamate and piso-metacarpal ligaments reach the hamate and fifth metacarpal, respectively (see Figure 6-3).[1-3]
Palmaris Longus This is a slender muscle, present less than 90% of the time, which resides on the medial side of the flexor carpi radialis muscle.[3] The muscle arises from the medial epicondyle via common flexor tendon and the intermuscular septum.	The muscle then travels distally the forearm to attach to the flexor retinaculum and the palmar aponeurosis. When present, it is the most superficial of the flexor tendons at the wrist.

Muscles in the Hand

Origin of the Hand Muscles	Insertion of the Hand Muscles
Abductor Pollicis Brevis (APB) This muscle arises from the trapezium and the flexor retinaculum (see Figure 6-4).	The APB will traverse itself and the fibers will attach to the base of the proximal phalanx on the radial side of the thumb (see Figure 6-4).[1,3,14]
Abductor Pollicis Longus (APL) This muscle arises from the dorsal aspect of the radius and ulna and the interosseus membrane (see Figure 6-4).	The muscle will spiral downward for its insertion and will attach itself on the radial side of the base of the first metacarpal on the palmar surface.[1,3,14]
Extensor Pollicis Brevis (EPB) This muscle arises from the dorsal surface of the radius and the interosseus membrane, just distal to the APL.	The EPB attaches to the dorsal surface of the proximal phalanx of the thumb. Together with APL, it forms the anterior border of the "anatomic snuff box."[1,3,11]
Extensor Pollicis Longus (EPL) This muscle arises from the posterior surface of the ulna near its mid-portion and also the interosseus membrane (see Figure 6-4).[1,3]	The EPL travels to the distal phalanx of the thumb and attaches via the extensor expansion. Its tendon forms the posterior border of the "anatomic snuff box" (see Figure 6-4).[1,14]

Pathomechanics of Common Injuries

The forearm muscles are prone to a variety of common maladies. One of the most common pathologies that can occur on either side of the elbow is **tendinitis**.[1,15,16]

The lateral side of the elbow will give rise to what is commonly known as "tennis elbow" or **lateral epicondylitis**. This pathology is common and will affect the muscles of the common extensor tendon, notably the extensor carpi radialis longus and brevis.[17-19] This condition can be brought on by repetitive motion in the workplace[20] or even from sporting activities.[21] On the medial side of the elbow, one of the most common pathologies is what is called "golfers elbow" or **medial epicondylitis**.[1,22] Wang[23] determined that part of the common flexor tendon is typically irritated during sporting activities such as throwing a ball. The common flexor tendon that is attached to the medial epicondyle can become inflamed with repetitive motion.[21,24-26]

A rare elbow problem that can occur on the lateral side of the forearm is called **radial tunnel syndrome** (RTS).[27,28] This can also be caused by repetitive motion resulting in the posterior interosseus branch of the radial nerve being trapped under the arcade of Froshe at the proximal opening of the supinator muscle in the forearm.[29-31] Henry and Stutz[32] stated that RTS is a particular condition that is different from tennis elbow. This is an important determination since many clinicians do not recognize RTS. Once the tennis elbow pain is resolved, there may still be some lingering pain in the forearm, usually palpable near the supinator muscle. In many cases the irritation to the radial nerve branch can be resolved with physical therapy.

Besides the tendinitis problems that occur at the elbow, other less common problems can involve fractures. The elbow can have fractures that involve the lateral epicondyle,[33] distal humerus,[34] and fracture dislocations of the radius and ulna.[35] The radial head can be fractured in direct trauma, such as a fall on an outstretched hand.[36] Olecranon fractures can occur as well.[37,38] Another fracture that occurs at the elbow is what is referred to as a Monteggia fracture, which involves a fractured ulna and a dislocated radial head.[1] This is another radial head fracture but involves the ulna as well.

One of the most common orthopedic problems that occur in the wrist deals with the tendons on the radial side of the hand. This is referred to as **DeQuervain's syndrome**.[1,39] This is a tenosynovitis that affects the first dorsal compartment of the "anatomic snuffbox." More specifically, the abductor pollicis longus and the extensor pollicis brevis tendon sheaths are inflamed in this syndrome.[39] Another common pathology that occurs in the wrist involves the flexor carpi ulnaris tendon. In many cases, this tendinitis comes from repetitive motion activities and also from racquet sports, which involve twisting the wrist and stressing the tendon attachment points.[1,40]

There are many other pathological conditions that can occur with the hand and wrist, but they are less common than the ones listed above. The extensor pollicis longus tendon can have tendinitis associated with it and occurs more commonly in patients who have rheumatoid arthritis.[1,41] Also, people who play the drums, using their wrist in an extended position, are prone to this type of ailment.

A more common ailment in the hand that involves tendons or the tendon sheath is **triggering.** This condition will typically occur on the flexor side of the hand and in the thumb, ring, or middle fingers.[1,42,43] The tendon sheath can become narrowed or was initially designed that way and through repetitive wear, the pulley (or sheath) becomes inflamed, leading to triggering, which is a catch in movement, sometimes followed by an audible pop as the tendon releases.[42,44,45] The specific reason patients wind up with this condition is unknown, but diabetics, menopausal females, and children seem predisposed to this condition.[1,43,44]

Clinical Pearls: Elbow, Wrist, and Hand

- Both the elbow and wrist are used in daily functional activities
- Forearm extensor muscles involved in tennis elbow from repetitive use
- Forearm flexor muscles involved with golfer's elbow
- Forearm flexors aid the bicep muscle with carrying objects
- Forearm extensor muscles are needed to support the lateral side of the elbow
- Carpal bones are attachment points for the forearm muscles, both the flexor and extensors
- Hand intrinsic muscles are needed for fine motor movements
- Triggering occurs in the thumb, index, or ring fingers
- Flexor carpi ulnaris tendinitis is common on the medial side of the elbow
- DeQuervain's syndrome: painful ulnar deviation, thumb flexion, and adduction
- Supination and pronation require lateral and medial gliding of the ulna on the humerus

References

1. Dutton, M. *Orthopaedic Examination, Evaluation, and Intervention.* New York: McGraw-Hill; 2008.
2. Tubiana R, Thomine MJ, Mackin E. *Evaluation of the Hand and Wrist.* 2nd ed. St. Louis, MO: Mosby; 1996.
3. Clemente CD. *Gray's Anatomy.* 30th ed. Philadelphia, PA: Lea and Febiger; 1985.
4. Daniels JM 2nd, Zook EG, et al. Hand and wrist injuries: part I. Nonemergent evaluation. *Am Fam Physician.* 2004;69(8):1941-1948.

5. Norkin CC, Levangie PK. *Joint Structure and Function: A Comprehensive Analysis.* 2nd ed. Philadelphia: FA Davis; 1992.

6. Keir PJ, Wells RP, Ranney DA. Passive properties of the forearm musculature with reference to hand and finger postures. *Clin Biomech (Bristol, Avon).* 1996;11(7):401-409.

7. Tay SC, van Riet R, et al. A method for in-vivo kinematic analysis of the forearm. *J Biomech.* 2008;41(1):56-62.

8. Neumann DA, Cook TM. Effect of load and carrying position on the electromyographic activity of the gluteus medius muscle during walking. *Phys Ther.* 1985;65(3):305-311.

9. Bremer AK, Sennwald GR, et al. Moment arms of forearm rotators. *Clin Biomech (Bristol, Avon).* 2006;21(7):683-691.

10. Galik K, Baratz ME, et al. The effect of the annular ligament on kinematics of the radial head. *J Hand Surg Am.* 2007;32(8):1218-1224.

11. Standring S. *Gray's Anatomy: The Anatomical Basis of Clinical Practice.* 40th ed. New York: Churchill Livingstone;, 2008.

12. Erak S, Day R, et al. The role of supinator in the pathogenesis of chronic lateral elbow pain: a biomechanical study. *J Hand Surg Br.* 2004;29(5):461-464.

13. Warfel J. *The Extremities: Muscles and Motor Points.* Philadelphia: Lea and Febiger; 1985.

14. Hollinshead WH, Rosse C. *Textbook of Anatomy.* 4th ed. Philadelphia: Harper & Row; 1985.

15. O'Dwyer KJ. Howie CR. Medial epicondylitis of the elbow. *Int Orthop.* 1995;19(2):69-71.

16. Maitland GD, Corrigan B. *Practical Orthopaedic Medicine.* London: Butterworths; 1987.

17. Rineer CA, Ruch DS. Elbow tendinopathy and tendon ruptures: epicondylitis, biceps and triceps ruptures. *J Hand Surg Am.* 2009;34(3):566-576.

18. Oron A, Schwarzkopf R, et al. Tennis elbow (lateral epicondylitis)—assessment and treatment. *Harefuah.* 2008;147(4):340-343, 373.

19. Szabo SJ, Savoie FH 3rd, et al. Tendinosis of the extensor carpi radialis brevis: an evaluation of three methods of operative treatment. *J Shoulder Elbow Surg.* 2006;15(6):721-727.

20. Fan ZJ, Silverstein BA, et al. Quantitative exposure-response relations between physical workload and prevalence of lateral epicondylitis in a working population. *Am J Ind Med.* 2009;52(6):479-490.

21. Hume PA, Reid D, et al. Epicondylar injury in sport: epidemiology, type, mechanisms, assessment, management and prevention. *Sports Med.* 2006;36(2):151-170.

22. Shiri R, Viikari-Juntura E, et al. Prevalence and determinants of lateral and medial epicondylitis: a population study. *Am J Epidemiol.* 2006;164(11):1065-1074.

23. Wang Q. Baseball and softball injuries. *Curr Sports Med Rep.* 2006;5(3):115-119.

24. Chen FS, Rokito AS, et al. Medial elbow problems in the overhead-throwing athlete. *J Am Acad Orthop Surg.* 2001;9(2):99-113.

25. Hotchkiss RN. Epicondylitis—lateral and medial. A problem-oriented approach. *Hand Clin.* 2000;16(3):505-508.

26. Field LD, Savoie FH. Common elbow injuries in sport. *Sports Med.* 1998;26(3):193-205.

27. Smola C. About the problem of radial tunnel syndrome or where does the tennis elbow end and where does the radial tunnel syndrome begin? *Handchir Mikrochir Plast Chir.* 2004;36(4):241-245.

28. Kleinert JM, Mehta S. Radial nerve entrapment. *Orthop Clin North Am.* 1996;27(2):305-315.

29. Clavert PJ, Lutz C, et al. Frohse's arcade is not the exclusive compression site of the radial nerve in its tunnel. *Orthop Traumatol Surg Res.* 2009;95(2):114-118.

30. Colbert SH, Mackinnon SE. Nerve compressions in the upper extremity. *Mo Med.* 2008;105(6):527-535.

31. Konjengbam M, Elangbam J. Radial nerve in the radial tunnel: anatomic sites of entrapment neuropathy. *Clin Anat.* 2004;17(1):21-25.

32. Henry M, Stutz C. A unified approach to radial tunnel syndrome and lateral tendinosis. *Tech Hand Up Extrem Surg.* 2006;10(4):200-205.

33. Rakonjac Z, Brdar R. Importance of initial fracture crack width in minimally dislocated fractures of humeral lateral condyle in children for evaluation of fracture stability and treatment choice. *Srp Arh Celok Lek.* 2009;137(3-4):179-184.

34. Sanchez-Sotelo J. Distal humeral nonunion. *Instr Course Lect.* 2009;58:541-548.

35. Verettas DA, Drosos GI, et al. Simultaneous dislocation of the radial head and distal radio-ulnar joint. A case report. *Int J Med Sci.* 2008;5(5):292-294.

36. Beingessner DM, Dunning CE, et al. The effect of radial head fracture size on radiocapitellar joint stability. *Clin Biomech (Bristol, Avon).* 2003;18(7):677-681.

37. Newman SD, Mauffrey C, et al. Olecranon fractures. *Injury.* 2009;40(6):575-581.

38. Sahajpal D, Wright TW. Proximal ulna fractures. *J Hand Surg Am.* 2009;34(2):357-362.

39. Schned ES. DeQuervain tenosynovitis in pregnant and postpartum women. *Obstet Gynecol.* 1986;68(3):411-414.

40. Helal B. Chronic overuse injuries of the piso-triquetral joint in racquet game players. *Br J Sports Med.* 1978;12(4):195-198.

41. Parellada AJ, Gopez AG, et al. Distal intersection tenosynovitis of the wrist: a lesser-known extensor tendinopathy with characteristic MR imaging features. *Skeletal Radiol.* 2007;36(3):203-208.

42. Nasca RJ. Trigger finger. A common hand problem. *J Ark Med Soc.* 1980;76(10):388-390.

43. Medl WT. Tendonitis, tenosynovitis, Trigger finger, and Quervain's disease. *Orthop Clin North Am.* 1970;1(2):375-382.

44. De la Parra-Marquez ML, Tamez-Cavazos TR, et al. Risk factors associated with trigger finger. Case-control study. *Cir Cir.* 2008;76(4):317-321.

45. Moore JS. Flexor tendon entrapment of the digits (trigger finger and trigger thumb). *J Occup Environ Med.* 2000;42(5):526-545.

Bibliography

Captier G, Canovas F, et al. Biometry of the radial head: biomechanical implications in pronation and supination. *Surg Radiol Anat.* 2002;24(5):295-301.

.

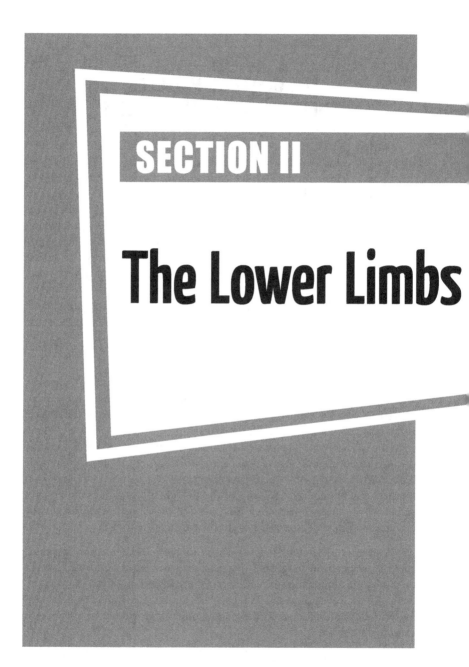

SECTION II

The Lower Limbs

Benjamin S, Bechtel RH, Conroy VM.
Cram Session in Functional Anatomy:
A Handbook for Students & Clinicians (pp. 45-86)
© 2011 Taylor & Francis Group.

7

THE GLUTEAL MUSCLES:
THEIR ROLE AND FUNCTIONS

During the US victory over the Russian hockey team in the 1980 Olympics, there were lots of bumps and bruises. There were a lot of hits from the US hockey team on the Russian team and vice versa. The hits were made via the shoulders pads, the elbows, and even the gluteal pads that were in the hockey pants. The gluteals muscles are needed for skating and also for providing the crushing hits that set the tone for one of the most amazing victories in sports history. Various studies have demonstrated the importance of gluteal function for specific sports and injuries can occur during either collegiate or recreational activities when these muscles do not function optimally.[1-4] The gluteal muscles are just as important for us to stand up from a chair, walk, run, or dance. Their optimal performance is imperative for hip stability that is required for us to function on a daily basis.[5]

Function of the Gluteal Muscles

The gluteal muscles are composed of the gluteus maximus, medius, and minimus (Figures 7-1, 7-2, and 7-3).[5] The gluteus maximus is the most superficial of all 3 gluteal muscles (see Figure 7-2). The size of the muscle is the muscle's trademark and it works as a main extensor of the hip along with the hamstrings.[5] The powerful gluteus maximus is used in a dual role during the gait cycle.[6,7] As we walk, the lower fibers of the gluteus maximus are active at initial contact, acting as a hip extensor to counteract the flexion moment due to impact.[7,8] The superficial portion of the gluteus maximus is active during mid-stance, when it is called upon to work with the gluteus medius as a hip abductor to maintain the trunk upright.[7,9]

Amaro et al[10] found that weakness in the gluteus medius/minimus, such as can occur in conditions like osteoarthritis, has a significant impact on a person's ability to walk with normal speed and quality of movement. Gluteal muscle weakness has also been shown to arise from sacroiliac joint mechanical dysfunctions, so these joints must be included in a thorough examination of the hip.[11]

Besides the gait cycle and daily functional activities, the gluteals are a focus during rehabilitation.[12] Recently is has been demonstrated that the gluteal muscles are an important link in the vestibular system's control of posture and maintaining balance.[13] Joseph et al[14] showed that muscle

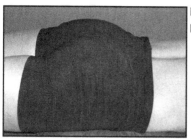

Figure 7-1A. The gluteal muscles relaxed in the prone position.

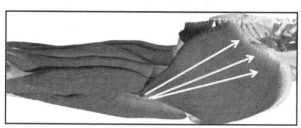

Figure 7-1B. Schematic of the gluteal muscles with forces shown by the arrows from the insertion point. (Reprinted with permission from Primal Pictures, 2009.).

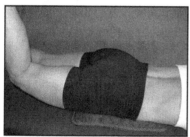

Figure 7-2A. The gluteal muscles contracted in vivo in prone.

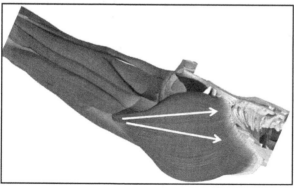

Figure 7-2B. Internal schematic of the gluteal muscles with forces in arrows from origin to insertion. (Reprinted with permission from Primal Pictures, 2009.)

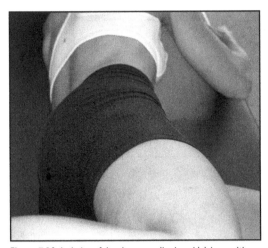

Figure 7-3A. Isolation of the gluteus medius in a sidelying position.

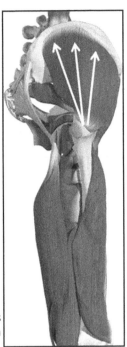

Figure 7-3B. The gluteus medius in a schematic and also the muscle forces shown by the arrows from the insertion point. (Reprinted with permission from Primal Pictures, 2009.)

imbalance involving a weak gluteus maximus and overcompensation from the gluteus medius and minimus results in internal rotation of the femur, which will in turn cause a dysfunctional gait pattern, with poor pelvic control and increased energy cost of walking. From a rehabilitation perspective, the clinician needs to address the gluteal tightness as well as the potential inhibition of the gluteus maximus, which can arise from dysfunction in the sacroiliac joint, the hip, or the lumbar spine.

Origin and Insertion

Origin of the Gluteal Muscles	Insertion of the Gluteal Muscles
GLUTEUS MEDIUS: The gluteus medius is the largest of the hip abductors and takes origin from the outer surface of the iliac crest, the posterior gluteal above and the anterior gluteal line below (see Figure 7-3).	Gluteus medius inserts on the lateral surface of the greater trochanter of the femur (see Figure 7-3).[5,15,16]
GLUTEUS MINIMUS: The gluteus minimus arises from the iliac crest, between the anterior and inferior gluteal lines.	The gluteus minimus will make its way to the anterior aspect of the greater trochanter for its insertion.[5,15,16]

(Continued)

Origin of the Gluteal Muscles	Insertion of the Gluteal Muscles
GLUTEUS MAXIMUS: The gluteus maximus arises from the posterior gluteal line, the lower lateral sacrum and coccyx bone, the fascia of the thoracodorsal and also the aponeuroses is the erector spinalis.[16,17] The muscle will also arise from the sacrotuberous ligament (see Figure 7-2).	About two-thirds of the fibers, the more superficial, traverse downward and obliquely laterally and cross the greater trochanter to attach to the iliotibial band. One-third of the fibers, the deepest, attach directly to the gluteal tuberosity on the posterior femur (see Figure 7-2).[5,15-17]
ACTION OF THE GLUTEAL MUSCLES	
The gluteus medius and the minimus will act primarily as abductors during the gait cycle.[18,19] They will be active in concert with the gluteus maximus at initial contact and during mid stance.[7,20-22] A majority of the fibers of gluteus maximus insert into the iliotibial band where they can aid in hip abduction but the deeper fibers insert directly into the femur and act to extend the hip, but under certain conditions can actually adduct the thigh at the hip joint.[16,17]	

Pathomechanics of Common Injuries

The gluteal muscles are situated in a place where direct trauma is not common, but injuries do occur to this muscle group.[5] The gluteus medius and maximus are prone to **tendinitis**[5,23] and this can cause pain when a person is walking, running, and going up and down stairs. With the gluteus medius' tendon inserting on the greater trochanter[17] this can leave the bone subject to injury. Injuries to the trochanteric bursae can occur from contact with the gluteus medius tendon or by direct trauma, such as a blow to the bursa.[5,23]

A fairly common injury in athletes involves the gluteus maximus attachment to the **iliotibial band,** which can lead to iliotibial band friction syndrome (see Figure 7-2).[2] This involves pain along the lateral portion of the thigh[16] and is frequently seen when a runner has poor balance between the muscles that surround the pelvis.[2]

When individuals have a total hip replacement, their gluteal muscles are many times affected by the surgical procedure either due to direct trauma to the muscles or from incidental damage to the muscles' nerve supply.[24,25] When that occurs, postsurgically a person will exhibit gait abnormalities and have loss of function. Many patients, on the other hand, can have weakness before surgery due to pain inhibition, and then afterward they may experience a further decrement, making it harder for therapists to strengthen the supporting hip muscles.[24,26] Despite those problematic issues, clinicians will need to focus on the correct exercises (against-gravity and gravity-minimized activities) to help those patients once they are out of surgery. Lastly, besides bursitis, tendinitis, and total hips, the gluteal muscles can be specifically affected if a person sustains a **fracture** of the hip.[27,28] Gluteal muscle injury will affect his or her **gait**[7] and also muscle tone. It is very important once again to outline the proper exercises needed to help these patients so that they can recover without a lot of atrophy and resulting gait abnormalities.

Clinical Pearls: Gluteal Muscles

- The gluteals are needed for pelvic stability for daily activities and sports

- The gluteals are needed for proper gait cycle functioning

- The gluteals are needed for supporting the hip and thigh when walking, getting up from a chair, and playing sports

- The gluteals will aid stair climbing and during functional activities

- The gluteals will indirectly support the lumbar spine via the stability of the muscles around the pelvis and hips

- The gluteals will affect the sacroiliac joint and mobility within that joint

References

1. Geraci MC Jr, Brown W. Evidence-based treatment of hip and pelvic injuries in runners. *Phys Med Rehabil Clin N Am.* 2005;16(3):711-747.
2. Fredericson M, Wolf C. Iliotibial band syndrome in runners: innovations in treatment. *Sports Med.* 2005;35(5):451-459.
3. Sloniger MA, Cureton KJ, et al. Lower extremity muscle activation during horizontal and uphill running. *J Appl Physiol.* 1997;83(6):2073-2079.
4. Shaffer BF, Jobe W, et al. Baseball batting. An electromyographic study. *Clin Orthop Relat Res.* 1993;(292):285-293.
5. Dutton M. *Orthopaedic Examination, Evaluation, and Intervention.* New York: McGraw-Hill; 2008.
6. Al-Hayani A. The functional anatomy of hip abductors. *Folia Morphol (Warsz).* 2009;68(2):98-103.
7. Perry J. *Gait Analysis: Normal and Pathological Function.* Thorofare, NJ: SLACK Incorporated; 1992.
8. Arnold AS, Komattu AV, et al. Internal rotation gait: a compensatory mechanism to restore abduction capacity decreased by bone deformity. *Dev Med Child Neurol.* 1997;39(1):40-44.
9. Preece SJ, Graham-Smith P, et al. The influence of gluteus maximus on transverse plane tibial rotation. *Gait Posture.* 2008;27(4):616-621.
10. Amaro A, Amado F, et al. Gluteus medius muscle atrophy is related to contralateral and ipsilateral hip joint osteoarthritis. *Int J Sports Med.* 2007;28(12):1035-1039.
11. Dorman T, Brierly S, Fray J, Pappani K. Muscles and the pelvic clutch: hip abductor inhibition in anterior rotation of the ilium. *Journal of Orthopedic Medicine.* 1995;17(3):96-100.
12. Distefano LJ, Blackburn JT, et al. Gluteal muscle activation during common therapeutic exercises. *J Orthop Sports Phys Ther.* 2009;39(7):532-540.
13. Strang AJ, Choi HJ, et al. The effect of exhausting aerobic exercise on the timing of anticipatory postural adjustments. *J Sports Med Phys Fitness.* 2008;48(1):9-16.

14. Joseph B. Treatment of internal rotation gait due to gluteus medius and minimus over-activity in cerebral palsy: anatomical rationale of a new surgical procedure and prelimi-nary results in twelve hips. *Clin Anat.* 1998;11(1):22-28.

15. Warfel J. *The Extremities: Muscles and Motor Points.* Philadelphia: Lea and Febiger; 1985.

16. Clemente CD. *Gray's Anatomy.* 30th ed. Philadelphia: Lea and Febiger; 1985.

17. Standring S. *Gray's Anatomy: The Anatomical Basis of Clinical Practice.* 40th ed. New York: Churchill Livingstone; 2008.

18. Oi N, Iwaya T, et al. FDG-PET imaging of lower extremity muscular activity during level walking. *J Orthop Sci.* 2003;8(1):55-61.

19. Bechtol CO. Correction of gluteus medius gait: dynamic balance gait. *Am Surg.* 1959;25:847-849.

20. Anderson FC, Pandy MG. Individual muscle contributions to support in normal walk-ing. *Gait Posture.* 2003;7(2):159-169.

21. Lyons KJ, Perry J, et al. Timing and relative intensity of hip extensor and abductor muscle action during level and stair ambulation. An EMG study. *Phys Ther.* 1983;63(10):1597-1605.

22. Bullock-Saxton JE, Janda V, et al. Reflex activation of gluteal muscles in walking. An approach to restoration of muscle function for patients with low-back pain. *Spine.* 1993;18(6):704-708.

23. Maitland GD, Corrigan B. *Practical Orthopaedic Medicine.* London: Butterworths; 1987.

24. Roth A, Venbrocks RA. Total hip replacement through a minimally invasive, anterolateral approach with the patient supine. *Oper Orthop Traumatol.* 2007;19(5-6):442-457.

25. Perves A, Perreau M, et al. Gluteal posterior approach for surgery of the acetabulum. *Rev Chir Orthop Reparatrice Appar Mot.* 1995;81(7):639-642.

26. Bell SN. Trans-gluteal approach for hemiarthroplasty of the hip. *Arch Orthop Trauma Surg.* 1985;104(2):109-112.

27. Sato Y, Inose M, et al. Changes in the supporting muscles of the fractured hip in elderly women. *Bone.* 2002;30(1):325-330.

28. Braun W, Mayr E, et al. Reconstruction of complex acetabular fractures using the extensile kocher-langenbeck approach (modified Maryland approach). *Oper Orthop Traumatol.* 1997;9(2):83-96.

8

THE HIP ABDUCTORS, ADDUCTORS, AND ROTATORS

As you move during your daily life, there is probably little effort when going up and down the stairs, shifting from side to side, getting in and out of bed, or even enjoying sporting activities. The motions that we make during our various activities usually happen unconsciously, and we give little thought to the complex processes that occur in the background to enable us to perform these intricate motor programs. In this chapter, we discuss the hip and some of the muscles that control the hip. We will also discuss some pathomechanical issues that may arise if one or more of the muscles are adversely affected.

Function of the Hip Abductors

The pelvis is made up of 2 innominate (hip) bones as well as the sacrum and the coccyx.[1] On the inner side of the innominates lie the sacroiliac joints, formed between innominates and the first 3 sacral vertebrae. On the outer side of the innominates, anterior to the sacroiliac joints, are the hip joints, formed between the concave acetabula on the innominates and the heads of the femurs. With this configuration, forces and moments acting across the hip also affect the sacroiliac joint and vice versa. The hip joint is designed, unlike the shoulder, to emphasize stability over mobility, and the acetabulum is designed to be deep to hold the femoral head in tightly. Surrounding the hip joint is what is referred to as the acetabular labrum, a fibrocartilaginous ring that further deepens the hip socket. Finally, the tubular hip capsule joins the acetabulum and the femoral neck like a Chinese finger puzzle, adding to stability in the joint.[2,3] As we move on to the muscles of the hip, the main hip abductors consist of the tensor fascia lata, the gluteus medius, and the gluteus minimus (Figure 8-1). These are aided by the secondary abductors, gluteus maximus, piriformis, and the superior and inferior gemelli.

The main muscle within the group that produces these actions is the gluteus medius (see Figure 8-1). The abductor group is needed during functional activities like walking, especially during the end of swing to the mid-stance phase of the cycle, when they hold us upright against gravity.[4] The gluteus medius and the minimus are the most active abductors in the group and with the leg fixed, help to keep the trunk and pelvis level as we transition to single limb stance at the beginning of the gait cycle and through stance phase.[4]

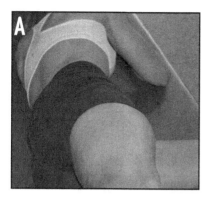

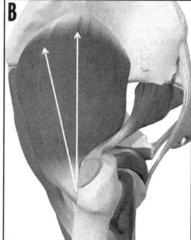

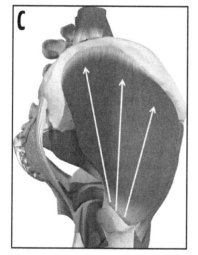

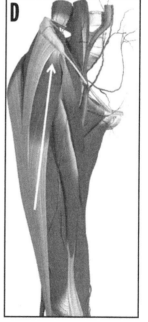

Figure 8-1. (A) Abductor group sidelying. Schematic of the gluteus minimus (B) and medius (C) and the forces from the insertion to the origin. (D) Schematic of the tensor fascia lata showing forces and vector from the iliotibial band to the hip. (Reprinted with permission from Primal Pictures, 2009.)

Origin and Insertion

Origin of the Hip Abductor Muscles	Insertion of the Hip Abductor Muscles
Tensor Fascia Lata (TFL): This muscle arises from the anterior outer portion of the iliac crest and the anterior border of the ilium (ASIS) (see Figure 8-1).	The TFL travels downward and backward to attach to the iliotibial tract (see Figure 8-1).[2,5,6]
Gluteus Medius: The gluteus medius is a prime hip abductor and takes origin from the outer surface of the iliac crest, between the posterior gluteal line and the anterior gluteal line (see Figure 8-1).	Gluteus medius inserts on the lateral surface of the greater trochanter (see Figure 8-1).[2,5,6]
Gluteus Minimus: The gluteus minimus arises from the iliac crest, the anterior and inferior gluteal lines.	The gluteus minimus will make its way to the anterior aspect of the greater trochanter for its insertion (see Figure 8-1).[2,5,6]
Action of the Hip Abductors	
The abductor muscle group causes the femur to move away from the mid-line of the body if the leg is free, or if the leg is fixed, causes the pelvis to tilt toward the active side.[2,7]	

Pathomechanics of Common Injuries

The hip abductors are a much needed muscle group for daily life activities. They are mostly needed for walking and other functions as mentioned previously. The hip abductors are important for not only level walking but for stair climbing as well,[8] and the powerful gluteus medius is needed to keep the pelvis and trunk level during this functional activity.[4,8]

If a person has an injury, muscle pathology, or neurological problem (eg, a stroke) that results in weakness of the gluteus medius, he or she may exhibit a **Trendelenberg gait**.[9] In a Trendelenberg gait pattern, the person will lean away from the weak side to "hang" on the stretched ligaments and joint capsule. There is also a **compensated Trendelenberg gait**, in which the person leans toward the affected hip as a way of minimizing the adductor moment about the affected hip due to gravity. Either of these gaits increases lateral movement during walking, which is a characteristic sign.[4,10]

When a person is walking and the nonsupporting limb is swinging forward, the gluteus medius on the supporting side keeps the pelvis from collapsing laterally into adduction at the supporting hip. Therefore, if the right leg is moving forward, the left gluteus medius is contracting to support the right side of the pelvis.[4,5,11] Henriksen et al[12] reported that if the gluteus medius is reduced in function, a person can develop **knee pain** due to increased stress, which could eventually lead to **osteoarthritic** changes.[13] The abductors specifically have such an important role in hip joint function that if the abductors have a reduction in their abilities and atrophy ensues in the muscle, the ipsilateral hip, and even the contralateral hip can develop **osteoarthritis**.[14] Another common ailment for the hip abductor group is **trochanteric bursitis**, or **tendinitis**. As it attaches to the greater trochanter, the gluteus medius can become inflamed[15] and cause pain in the hip, low back, and even give symptoms mimicking sciatica.[2,16-18]

Clinical Pearls: Hip Abductors

- Needed for hip stability walking and standing

- Needed for knee stability and to prevent knee injury and biomechanical changes

- Needed during the gait cycle during stance and swing phases

- Linked to lower back pain and needed to support the lower back

- Reduced pressure on the knee and hip when functioning correctly

- Needed for inclines and stair walking to control pelvic stability

The Hip Adductors

Function and Functional Anatomy

The adductor muscles are on the medial side of the hip joint.[2,5] The adductor group consists of the adductor magnus, the adductor longus, and the gracilis muscle (Figure 8-2). Beside the main 3 adductor muscles, the adductor brevis and the pectineus are secondary adductors on the higher portion and the medial side of the joint. The pectineus is deeply placed and forms the floor of the femoral triangle, through which the femoral artery, nerve, and lymphatics course to the thigh.

The adductor muscle group is active during the gait cycle.[4] The most important 3 adductors (adductor magnus, longus, and gracilis) are active from terminal stance to the mid-swing phase, helping to initiate hip flexion and maintain straight line motion.[4,19] The adductor magnus also shows EMG activity during the terminal phase of swing, where the vertical head acts like the hamstrings to limit flexion and the oblique head acts like an adductor to help bring the limb back into the midline.[4]

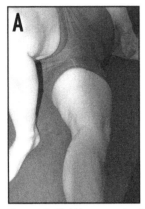
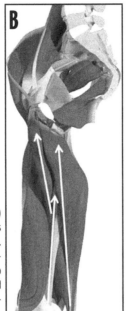
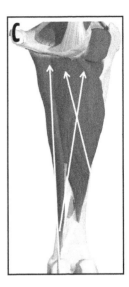

Figure 8-2. (A) Adductor muscles. (B) Schematic of the hip adductor muscles with muscle forces noted by arrows. (C) The gracilis and the adductor magnus with their forces from insertion to origin. The left arrow is the gracilis and the 2 crossing arrows are the adductor magnus with the 1 arrow coming from the adductor tubercle. (Reprinted with permission from Primal Pictures, 2009.)

Origin and Insertion

Origin of the Hip Adductor Muscles	Insertion of the Hip Adductor Muscles
ADDUCTOR MAGNUS: This is a large muscle that is triangular-shaped, with 2 heads of origin. The oblique head, sometimes called the adductor portion, originates from the ischio-pubic ramus, while the vertical head, sometimes called the hamstring portion, originates from the ischium and from the inferior and lateral portions of the ischial tuberosity along with the hamstrings (see Figure 8-2).[3,5]	From its origin the adductor magnus muscle travels downward and obliquely to attach to the medial femur along most of its length, spanning from the greater trochanter along the linea aspera to the adductor tubercle (see Figure 8-2).[2,5,6]
ADDUCTOR LONGUS: This muscle is the most anterior of all the adductor muscle that we will discuss. It arises from the anterior surface of the pubic bone and also from the pubic crest and symphysis (see Figure 8-2).[2,5]	The muscle travels a relatively short, diagonal course from its origin on the pubic bone to insert on the medial lip of the linea aspera on the femur (see Figure 8-2).[5,6]

(Continued)

Origin of the Hip Adductor Muscles	Insertion of the Hip Adductor Muscles
GRACILIS: This muscle is superficial and easily palpated on the medial thigh. It is broad at the top and tapers downward toward its insertion point. Gracilis takes origin from a thin insertion on the pubis near the symphysis pubis	The gracilis travels downward along the medial femur over and around the medial femoral condyle and the medial condyle of the tibia to attach to the medial tibia on the tibial crest.[2,5,6] The tendon of gracilis merges together with tendons from sartorius and semitendinosus just above their insertion, forming the common pes anserine tendon, which overlies the pes anserine bursa (see Figure 8-2).[3,5]
ACTION OF THE HIP ADDUCTOR MUSCLES	
The main function of the adductor group, with the leg free, is moving the leg toward midline, across the body. These muscles also aid in flexion and medial rotation of the thigh. With the leg fixed, these muscles can help with weight shifting that occurs in walking, by pulling the pelvis toward the supporting (stance) leg.[2,5,6]	

Pathomechanics of Common Injuries

The adductor muscle group is not frequently injured in daily activities but is prone to **strains and sprains** in certain sports. Because it crosses both the knee and hip, the gracilis is maximally stretched in a position of knee extension and hip abduction.[20,21] When this occurs, there can be injuries to the adductor muscle group and there also can be negative consequences for the hip labrum.[22,23] One of the most common hip injuries that can occur during sports is the **adductor strain,**[23] and it is common for hockey players to have this due to the nature of the skating stride, where the muscle is being stretched while contracting eccentrically.[24] Tyler et al[25] studied the prevalence of **hip adductor injuries** in National Hockey League players and concluded that training this muscle group could help in the long run to prevent further injury. Recently, it has further been shown that there is a strong relationship between poor adductor strength and abdominal muscle tone that can lead to sports hernias.[26]

Clinical Pearls: Hip Adductors

- Needed for proper gait sequencing

- Needed to support the medial side of the hip joint

- Common sports injury (strengthening needed), especially in hockey

- Needed to strengthen after fractures and total hip arthroplasty surgery for proper daily activities (walking, transitions)

- Can be painful during sports when running, cutting, jumping, and skating

- Pain common during sports hernias, but may also be due to muscle imbalance or sacroiliac joint dysfunction.[4,5,20,26,27]

- Function: Adductors will cause the thigh to adduct and can also flex the thigh at the hip

External Rotators of the Hip
Function and Functional Anatomy

This section will outline the 6 external rotators and their function. The external rotators of the hip (piriformis, obturator internus, obturator externus, superior gemelli, inferior gemelli, and the quadratus femoris) are named for their location and function (Figures 8-3A and B). The functional role of these muscles can change during movement. For example, the piriformis and the obturator internus are both lateral rotators of the thigh but it has been reported that when the hip is flexed beyond 90 degrees, the muscles increase their ability to abduct the thigh in addition to lateral rotation.[2,5]

Origin and Insertion

Origin of the Hip External Rotators	Insertion of the Hip External Rotators
PIRIFORMIS: This muscle is a pyramidal, flat muscle that takes origin from the pelvic surface of the sacrum between the first through fourth sacral foramina, from the anterior capsule of the sacroiliac joint and also the sacrotuberous ligament (Figure 8-3C).	The muscle emerges through the greater sciatic foramen just superficial to the sciatic nerve and travels forward and outward at about a 45-degree angle to reach the superior border of the greater trochanter (see Figure 8-3).[2,3,5]
OBTURATOR INTERNUS: This muscle is situated posterior to the hip joint and arises from above the inner surface of the innominate bone around the obturator foramen and from the obturator membrane (Figure 8-3D).	The muscle will travel from its attachment, over the top of the ischial tuberosity, joining between the superior and inferior gemelli as it does so, to reach the anterior portion of the medial side of the greater trochanter (see Figure 8-3).[2,5,6]
OBTURATOR EXTERNUS: This muscle is a flat, triangular muscle that arises from the outer surface of the innominate bone around the obturator foramen and from the medial two-thirds of the obturator membrane and the ramus of the pubis (Figure 8-3F).[5]	The muscle travels from its pelvic insertion to course laterally and attach to the trochanteric fossa of the greater trochanter of the femur (see Figure 8-3).[2,5,6,28]
SUPERIOR GEMELLUS (SG): This is a small muscle that arises from the ischial spine and blends in with the obturator internus muscle (Figure 8-3E).	The superior gemelli travel to the greater trochanter on the medial side and attach along with the tendon of the obturator internus.[5,6]

(Continued)

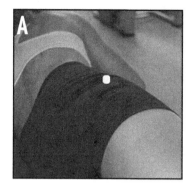
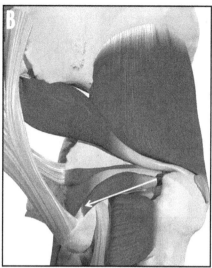
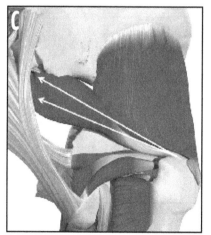
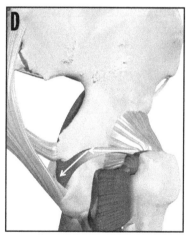

Figure 8-3. (A) The external rotators from an oblique angle in sidelying with the greater trochanter marked where the external rotators are inserted. (B through G) Schematics of the 6 external rotators of the hip. (B) The inferior gemelli. (C) Piriformis muscle. (D) Obturator internus. (Reprinted with permission from Primal Pictures, 2009.)

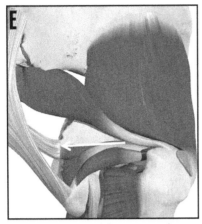

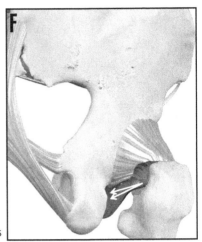

Figure 8-3. Schematics of the 6 external rotators of the hip. (E) The superior gemelli. (F) Obturator externus. (G) Quadratus femoris. Muscle forces shown by the arrows. (Reprinted with permission from Primal Pictures, 2009.)

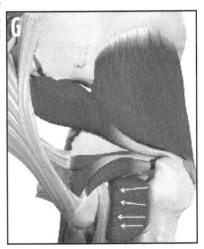

Origin of the Hip External Rotators	Insertion of the Hip External Rotators
Inferior Gemellus (IG): The larger of the 2 gemelli muscles, the IG arises from the lateral aspect of the ischial tuberosity and parallels the course of the obturator internus muscle, joining with that tendon and the tendon of superior gemellus (see Figure 8-3B).	The tendons of the SG and IG will join and travel to insert on the medial portion of the greater trochanter.

(Continued)

Origin of the Hip External Rotators	Insertion of the Hip External Rotators
QUADRATUS FEMORIS (QF): This is a short, flat, broad muscle, quadrilateral in size, situated below the obturator internus and superior and inferior gemelli muscles.[5] The QF arises from the ischial tuberosity on its lateral margin and curves laterally around the femoral neck to reach the posterior aspect of the femur (Figure 8-3G).	The QF will attach to a small tubercle on the intertrochanteric crest called the quadrate tubercle, and along the quadrate line toward the lesser trochanter (see Figure 8-3).[5,6]
ACTION OF THE HIP EXTERNAL ROTATORS	
The hip external rotators have a primary functional role to stabilize the hip and pelvis with the leg fixed, but they will also assist with hip abduction when the leg is free to move.[2,5,29]	

Pathomechanics of Common Injuries

With the possible exception of piriformis, the external rotators of the hip do not exhibit as high a prevalence of injury as the gluteal muscles or abductors do. From a clinical perspective, this means the focus can be more on strengthening this muscle group and less on repair and re-establishing motor control patterns.

The external rotators may be injured during certain surgical procedures (ie, **total hip arthroplasty**).[30,31] During this procedure, the rotators can be incised or reflected (the piriformis and the obturator externus),[31] which can lead to a decrement in strength output but not necessarily a poor prognosis for hip rehabilitation after such a surgery. Conversely, Goldstein et al[32] went on to conclude that to keep the stability in the hip joint and help prevent hip dislocation, the external rotators should be approximated by the surgeon to reduce the effects of postoperative inhibition. The external hip rotators were also shown to be important to support the pelvis and decrease stress on related joints. This can reduce the incidence of **knee pain**[13,33] whether in a couch potato or an athlete.[29] The external rotators are needed for hip stability, especially when walking, running, and competing. The patient who has a total hip arthroplasty will need to rehabilitate this muscle group to re-establish his or her strength and stability for functional activities.[34]

Like most of the lower extremity muscles, keep in mind that these muscles function when the foot is fixed on the ground in a closed kinetic chain. Under those conditions, piriformis, because of its origin on the sacrum, and the anteriorly oriented direction of its fibers, can pull the sacrum anteriorly (nutation). If this occurs on one side only, as in stair climbing or running, it can result in a sacral torsion, which creates a **hypomobile sacroiliac**

joint on the affected side. Thus, in patients with low back pain, it is always important to check hip motions including rotation. In particular, limited hip internal rotation on one side can indicate sacroiliac joint involvement due to a "tight" piriformis.

Clinical Pearls: Hip External Rotators

■ Piriformis pain may indicate sacroiliac joint involvement due to tension in the muscle

■ Needed for hip stability during closed chain activities and will aid with keeping the hip joint stable

■ The external rotators of the hip will help patients with patellar femoral pain; the muscles will help to keep proper alignment so that the distal quad muscle functions properly and will reduce the stress on the knee joint

■ The hip external rotators are needed for patients who have undergone total hip arthroplasty; this will give the patient stability and strength to aid with stair climbing and walking on level surfaces

References

1. Howie JL. Computed tomography of the bony pelvis: a protocol for multiplanar imaging. Part I: normal anatomy. *J Can Assoc Radiol.* 1985;36(4):278-286.
2. Dutton M. *Orthopaedic Examination, Evaluation, and Intervention.* New York: McGraw-Hill; 2008.
3. Hollinshead HW, Rosse C. *Textbook of Anatomy.* 4th ed. Philadelphia: Harper & Row; 1985.
4. Perry J. *Gait Analysis: Normal and Pathological Function.* Thorofare, NJ: SLACK Incorporated; 1992.
5. Clemente CD. *Gray's Anatomy.* 30th ed. Philadelphia: Lea and Febiger; 1985.
6. Warfel J. *The Extremities: Muscles and Motor Points.* Philadelphia: Lea and Febiger; 1985.
7. Norkin CC, Levangie PK. *Joint Structure and Function: A Comprehensive Anaylsis.* 2nd ed. Philadelphia: FA Davis Co; 1992.
8. Nadeau S, McFadyen BJ, et al. Frontal and sagittal plane analyses of the stair climbing task in healthy adults aged over 40 years: what are the challenges compared to level walking? *Clin Biomech (Bristol, Avon).* 2003;18(10):950-959.
9. Capello WN, Feinberg JR. Trochanteric excision following persistent nonunion of the greater trochanter. *Orthopedics.* 2008;31(7):711.
10. Liu MQ, Anderson FC, et al. Muscles that support the body also modulate forward progression during walking. *J Biomech.* 2006;39(14): 2623-2630.
11. Al-Hayani A. The functional anatomy of hip abductors. *Folia Morphol* (Warsz). 2009;68(2): 98-103.
12. Henriksen M, Aaboe J, et al. Experimentally reduced hip abductor function during walking: implications for knee joint loads. *J Biomech.* 2009;42:1236-1240.

13. Mascal CL, Landel R, et al. Management of patellofemoral pain targeting hip, pelvis, and trunk muscle function: 2 case reports. *J Orthop Sports Phys Ther.* 2003;33(11):647-660.

14. Amaro A, Amado F, et al. Gluteus medius muscle atrophy is related to contralateral and ipsilateral hip joint osteoarthritis. *Int J Sports Med.* 2007;28(12):1035-1039.

15. Walker P, Kannangara S. Lateral hip pain: does imaging predict response to localized injection? *Clin Orthop Relat Res.* 2007;457:144-149.

16. Farasyn AD, Meeusen R, et al. Validity of cross-friction algometry procedure in referred muscle pain syndromes: preliminary results of a new referred pain provocation technique with the aid of a Fischer pressure algometer in patients with nonspecific low back pain. *Clin J Pain.* 2008;24(5):456-462.

17. Bewyer DC, Bewyer KJ. Rationale for treatment of hip abductor pain syndrome. *Iowa Orthop J.* 2003;23:57-60.

18. Simons DG, Travell JG. Myofascial origins of low back pain. 3. Pelvic and lower extremity muscles. *Postgrad Med.* 1983;73(2):99-105, 108.

19. Rutherford DJ, Hubley-Kozey C. Explaining the hip adduction moment variability during gait: Implications for hip abductor strengthening. *Clin Biomech (Bristol, Avon).* 2009;24(3):267-273.

20. Tibor LM, Sekiya JK. Differential diagnosis of pain around the hip joint. *Arthroscopy.* 2008;24(12):1407-1421.

21. Braun P, Jensen S. Hip pain: a focus on the sporting population. *Aust Fam Physician.* 2007:36(6):406-408, 410-403.

22. Feeley BT, Powell JW, et al. Hip injuries and labral tears in the national football league. *Am J Sports Med.* 2008;36(11):2187-2195.

23. Morelli V, Smith V. Groin injuries in athletes. *Am Fam Physician.* 2001;64(8):1405-1414.

24. Daly PJ, Sim FH, et al. Ice hockey injuries. A review. *Sports Med.* 1990;10(2):122-131.

25. Tyler TF, Nicholas SJ, et al. The effectiveness of a preseason exercise program to prevent adductor muscle strains in professional ice hockey players. *Am J Sports Med.* 2002;30(5):680-683.

26. Caudill P, Nyland J, et al. Sports hernias: a systematic literature review. *Br J Sports Med.* 2008;42(12):954-964.

27. Standring S. *Gray's Anatomy: The Anatomical Basis of Clinical Practice.* 40th ed. New York: Churchill Livingstone; 2008.

28. Delp SL, Hess WE, et al. Variation of rotation moment arms with hip flexion. *J Biomech.* 1999;32(5):493-501.

29. Snyder KR, Earl JE, et al. Resistance training is accompanied by increases in hip strength and changes in lower extremity biomechanics during running. *Clin Biomech (Bristol, Avon).* 2009;24(1):26-34.

30. Curvale G, Rochwerger A, et al. Posterior position of acetabular component of a total hip prosthesis: possible cause of lower limb rotation problem. *Rev Chir Orthop Reparatrice Appar Mot.* 1998;84(7):653-656.

31. Stahelin T, Vienne P, et al. Failure of reinserted short external rotator muscles after total hip arthroplasty. *J Arthroplasty.* 2002;17(5):604-607.

32. Goldstein WM, Gleason TF, et al. Prevalence of dislocation after total hip arthroplasty through a posterolateral approach with partial capsulotomy and capsulorrhaphy. *J Bone Joint Surg Am.* 2001;83-A(Suppl 2, Pt 1):2-7.

33. Piva SR, Goodnite EA, et al. Strength around the hip and flexibility of soft tissues in individuals with and without patellofemoral pain syndrome. *J Orthop Sports Phys Ther.* 2005;35(12):793-801.

34. Yamaguchi T, Naito M, et al. The effect of posterolateral reconstruction on range of motion and muscle strength in total hip arthroplasty. *J Arthroplasty.* 2003;18(3):347-351.

$$\boxed{9}$$

THE QUADRICEPS AND HAMSTRINGS

The skater is crouched down, he moves from side to side with the puck, he pushes off to the side, and takes the puck into the corner. He gets checked by an opponent and is pushed down but gets up very quickly and passes the puck in front of the net to the next skater. The reason that the skater can perform those maneuvers, moving from side to side and also up and down, is not only due to the hip muscles that were discussed previously but also due to the quadriceps and hamstring complex. What is their role for support and function? This chapter will outline the quadriceps and hamstrings complex and discuss their function and pathology.

Function of the Quadriceps and Hamstrings

Almost everyone has heard of the "quads and hams." These muscles are very important to standing and walking upright. When we walk, move around, go up and down stairs, and get out of a chair, we use our quadriceps to help us initiate these motions. Both the hamstring and quadriceps are needed for an array of functional activities[1] and both work to move our legs. A surprising amount of the work of the quadriceps and hamstrings is done eccentrically, ie, the muscles generate force while lengthening, usually under the influence of gravity.[2,3] For example, when you are walking up and down the stairs, it is the role of the quadriceps and hamstrings to work together not only to push you up, but also control your decent eccentrically so you do not fall.[4,5] The quadriceps also has a major influence on patellar alignment at the knee, which is very important during sporting activities.[6,7]

During the gait cycle, the quadriceps are active (Figures 9-1A, 9-2A, and 9-2C).[1] The rectus femoris is active in the pre-swing phase and also at the initial swing phase (as a hip flexor).[1] Studies confirm that if you improve the quad power and function, gait will improve, especially in the elderly.[8] When you look at the all-important vastus medialis where studies show that is active mostly during the terminal swing phase and the initial contact of the foot on the ground (see Figure 9-2C).[1,9] During many upright activities in which we all participate on a daily basis, there is a need to have the quadriceps contract eccentrically, which will aid in limiting knee flexion (ie, during horseback riding).[10] The quadriceps is also needed during most sporting activities to provide concentric and eccentric control when the athlete is cutting, springing, riding, skating, or vaulting.[11-15]

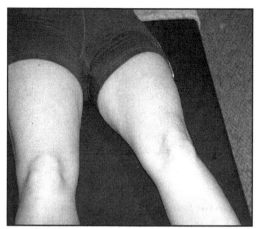

Figure 9-1A. Quadriceps muscle relaxed, both legs shown.

Figure 9-1B. Schematic of the right quadriceps with muscle forces as shown from insertion to origin as shown by the arrows. (Reprinted with permission from Primal Pictures, 2009.)

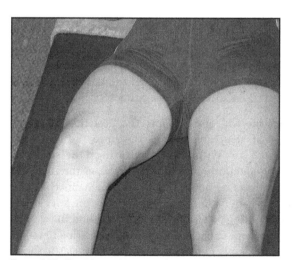

Figure 9-2A. The right quadriceps muscle contracted, with the vastus medialis and the vastus lateralis shown.

Figure 9-2B. Schematic shows the forces of the vastus medialis and the vastus lateralis with the rectus femoris reflected. (Reprinted with permission from Primal Pictures, 2009.)

Figure 9-2C. The right vastus medialis contracted and emphasized.

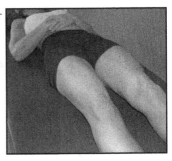

Now let's turn our attention to the hamstring muscles, which are composed of a 3-muscle system that is as important as the quadriceps to all the activities mentioned previously (Figures 9-3A and B).[1,16] Together as a group, all 3 hamstring muscles (semitendinosus, semimembranosus, and the bicep femoris) decelerate the swinging leg by generating hip extension and knee flexor moments during late and mid-swing.[1,9,17] During the walking cycle, for example, the hamstrings are needed to flex the knee, which occurs in the initial and mid-swing phases of the cycle (see Figure 9-3B).[1,18,19] During the initial loading phase of the foot, the hamstrings will be activated, but at a lower level than compared to what occurs during the swing phase. One interesting point that we can make here is that the semimembranosus and the semitendinosus are still active during mid-stance to support the medial side of the knee joint, which will compensate for the externally imposed valgus moment at the knee.[1,20] This will give support to the medial side of the knee during the gait cycle on and off uneven surfaces.

The hamstring muscles will also support the posterior aspect of the knee and the internal structures within the knee (ie, the anterior cruciate ligament [ACL]).[21] With strong hamstrings, the tibia can be prevented from gliding anteriorly and causing stress on the ACL.[21,22] In fact, in cases where patients do not have an ACL, the hamstrings are one of the focal points of rehabilitation in order to protect the knee internally and allow a normal walking pattern (see Figures 9-3A and B).[21,23] The hamstrings also have a role in lower back pain. Normal hamstring function helps to relieve stress on the lumbar spine by providing viscoelastic shock absorption. Stokes et al[24] showed that the attachment of the hamstring to the ischial tuberosity causes the pelvis to tilt in a posterior direction, which results in reducing the lumbar lordosis (see Figure 9-3C). Since the lordosis allows the spine to absorb shock and distribute load, if the hamstrings are too tight, some patients may experience low back pain.[24]

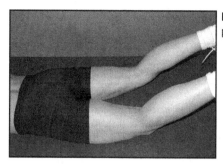

Figure 9-3A. The hamstrings contracted in the prone position.

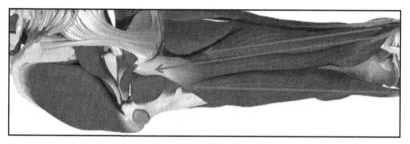

Figure 9-3B. The hamstrings with muscle forces as noted by the arrows. The arrows are shown from lateral to medial as the biceps femoris, semitendinosus, and the semimembranosus. (Reprinted with permission from Primal Pictures, 2009.)

In these days of evidence in rehabilitation and getting the most out of patients, many sporting activities require certain strength ratios between the hamstrings and the quadriceps (see Figures 9-2B and 9-3B). Clinicians should be aware that for most activities of daily life, the preferred ratio of quadriceps to hamstring strength is 1:1. This means that the hamstrings and quadriceps are equal in strength so that the knee joint is supported and will function properly.[25] By and large, the quadriceps muscle is usually stronger than the hamstrings, sometimes due to overtraining, but isokinetic evaluation reveals that the ratio is very close to one at slower speeds but at the higher speeds, the ratio is larger, thus showing that the quadriceps are better suited to rapid movements.[26-30] The stronger the hamstrings are to the quadriceps, the closer the ratio and the more stable the muscles will be around the knee. When certain injuries (ACL) occur, it is advantageous to have the ratio to be as close to 1:1 as possible. During rehabilitation or functional activities, the ACL will prevent the tibia from moving anteriorly, but also understand that the ratio can change as the knee motion changes.[31] Thus, as you flex or extend your knee, the primarily one-joint quadriceps muscles and the primarily multijoint hamstring muscles have to deal with different sets of demands. Overall, the closer to a 1:1 ratio you have, the more stability you will create within the knee joint.

Origin and Insertion

The Quadricep Muscles	
Origin of the Quadriceps Muscle	**Insertion of the Quadriceps Muscle**
RECTUS FEMORIS: The rectus femoris is in the middle of the thigh and is the only one of the 4 quadricep muscles that crosses the hip joint (see Figure 9-1A).[4,32] The rectus is a muscle that has 2 origination sites: the straight tendon arises from the anterior inferior iliac spine (AIIS) and the reflected tendon arises from the posterior aspect of the acetabular brim (see Figure 9-1B).[16,33]	Both tendons of the rectus femoris will travel together to their insertion site on the patella. The insertion will occur at a 5-degree angle,[4] that is to say the fiber orientation, like most of the quadriceps, is from lateral to medial as they traverse from hip to knee to attach to the superior portion of the patella and finally on the tibial tubercle (see Figures 9-1B and 9-2B).[4,16,33]
VASTUS LATERALIS: The vastus lateralis is the largest portion of the quadricep muscle[16] and will have a very broad origination. It arises from the capsule of the hip joint, intertrochanteric line of the femur, lateral portion of the gluteal line, and the anterior and posterior borders of the greater trochanter (see Figures 9-1B and 9-2A).[16,33,34]	The muscle travels downward from lateral to medial in an oblique fashion, at about a 12-degree angle toward the knee joint. Its tendon joins the others in the quadriceps expansion to attach to the lateral border of the patella and also the knee joint, with eventual attachment to the main patellar tendon at the tibial tubercle (see Figure 9-1B).[4,16,34]
VASTUS MEDIALIS: The vastus medialis is composed of 2 functional portions based on fiber directions, the vertical portion and the oblique portion[4,32] arising from the lower portion of the intertrochanteric line of the femur, the medial lip of the linea aspera, the tendons of the adductor longus and magnus (see Figures 9-2B and C).	The muscle travels downward to reach the patella and tibial tubercle. The oblique portion is unique among the quadriceps, since it represents the only component that can exert a medial pull on the patella to correct lateral tracking errors. The fibers of the oblique head of vatsus medialis attach at an angle of 45 degrees to the patella[4] and both the oblique and vertical heads attach to the medial side of the patella and the tibial tubercle (see Figure 9-2B).[16,33,34]
VASTUS INTERMEDIUS: This muscle lies in the midline deep to rectus femoris and arises from the upper two-thirds of the femur on the anterior and the lateral surface and also from the intermuscular septum. Its fiber orientation will be in line with the femur.[4]	The vastus intermedius will merge its fibers with the rectus femoris and the vastus muscles to attached on the patella and the tibial tuberacle.[4,16,34]
ACTION OF THE QUADRICEPS MUSCLE	
The most common functional activities of the quadriceps muscle are to extend the knee joint during daily and sporting activities.[5,16]	

The Hamstring Muscles

Origin of the Hamstring Muscles	Insertion of the Hamstring Muscles
BICEPS FEMORIS: The biceps femoris arises from the lateral and posterior aspects of the femur.[16,32] The muscle has 2 heads of origin (see Figures 9-3A and B). *THE LONG HEAD* originates from the ischial tuberosity in conjunction with the semitendinosus. The long head of the bicep femoris will also arise from portions of the sacrotuberous ligament. This part of the muscle crosses the hip joint. *THE SHORT HEAD* originates from the lateral lip of the linea aspera on the femur above the knee and the lateral intermuscular septum (see Figure 9-3B).[16,34]	The lateral head traverses obliquely downward to merge with fibers of the short head and together they arrive at the top of the fibular head where biceps femoris inserts (see Figure 9-3B).[33] A small part of the muscle also attaches to the lateral tibial condyle. An important point here is that the common fibular nerve will course on the medial side of the biceps femoris tendon as it curves around the fibular head (see Figure 9-3B).[16,34] The long head receives innervation from the tibial portion of the sciatic nerve and the short head is innervated by the common fibular nerve.
SEMITENDINOSUS: This muscle has a very long tendon that is palpable superficially on the posterior and medial aspects of the thigh.[16] The semitendinosus will arise from the ischial tuberosity on the inferomedial aspect[16,34] and joins the common hamstring tendon along with biceps femoris (see Figure 9-3B).	The muscle fibers end in about the middle of the thigh[16,33] and then the long tendon is formed. The tendon courses around the medial side of the tibial condyle and curves over the tibial collateral ligament to insert on the proximal portion of the tibia at the pes anserine insertion point along with sartorius and gracilis (see Figure 9-3B).[4,16,33]
SEMIMEMBRANOSUS: The semimembranosus is situated just deep to semitendinosus on the medial and posterior portion of the femur, but is easily palpable due to its wide tendon. Semimembranosus arises from the lateral and upper portions of the ischial tuberosity. The tendon attachment is wide and lies just medial to the common tendons for the bicep femoris and the semitendinosus (see Figure 9-3B).[4,16]	The muscle inserts into the medial tibial condyle but will have expanded fibrous insertions also attach laterally to the oblique popliteal ligament.[16] The tendon also has connections to the medial meniscus and will support the medial side of the knee joint.[4,16]
ACTION OF THE HAMSTRING MUSCLES	
The hamstrings as a collective group are needed to extend the thigh at the hip and flex the knee.[4,5]	

Pathomechanics of Common Injuries

There are many common injuries that can occur at the knee, but we will focus our attention on the quadriceps and hamstrings, respectively. The quadriceps as we discussed are composed of 4 muscles that will fill up the anterior compartment of the thigh. Common injuries that can occur to the quadriceps include **tendinitis of the patellar tendon** (jumper's knee),[35] and even **tendinitis that is proximal to the patella**.[35] In some cases, a person may experience low back pain, which can cause some inhibition to the quad muscle; even when the knee itself is injured that may lead to quadriceps dysfunction.[36] A noncommon problem in the knee that does occur under certain conditions is bone growth within the muscle, which is referred to as **myositis ossificans**. This may occur after trauma, such as a direct blow to the thigh, or as a result of systemic changes, such as accompany spinal cord injury.[37]

In the case where the patient experiences sporting injuries or household accidents, **patellofemoral subluxation** can occur. This can cause disabling pain and weakening of the quad muscle.[38-41] The quadriceps is also subject to other maladies that can occur with patellofemoral syndrome when the cartilage is worn and roughened under the patella, leading to the pain of **chondromalacia**.[40,42,43] Chondromalcia is treatable in a physical therapy setting, but care needs to be taken to avoid compression at the patellofemoral joint. Lastly, patients can also experience **rupture** of their **quadriceps tendon**, especially after a quick jerk or a fall. This situation causes an immediate release of the quadriceps tendon, probably resulting in a fall, and requires surgery to fix.[44]

In considering the patellofemoral joint, much of the focus is on the quadriceps mechanism and the lack of strength or strength imbalances between the vertical and oblique portions of vastus medialis of the patellofemoral mechanism.[39,42] The strength of the quadriceps is so important to daily life functioning and the vastus medialis is a crucial part of the quadriceps since its oblique fibers provide the only active component causing the patella to glide medially and avoiding lateral subluxation or dislocation. The oblique portion of the vastus medialis muscle resists the lateral pull of the other quadriceps muscles and allows the patella to track correctly. Poor tracking of the patellar can lead to knee pain and possible quadricep muscle inhibition.[4,36,45]

Common Hamstring Pathological Injuries

The hamstrings are subject to injury as the quadriceps are and some of the most common injuries include **hamstring tendinitis**[46,47] and **tendinosis** or overuse injury at the ischial attachment site.[48-50] A common injury that occurs to patients either after surgery or from a sports situation is hamstring weakness with resulting **poor quadricep/hamstring ratios to** protect the knee.[51-53] It is a common problem and it is very important to keep the hamstrings as strong as you possibly can so that more pressure is taken off of the ACL.

Other hamstring injuries that do occur but are not common are hamstring **rupture** at the proximal attachment[54-58] and **distal avulsion** of the hamstring tendon.[59,60] If a patient has a traumatic injury or surgery involving the hamstrings, he or she sometimes can present with **foot drop** due to involvement of the fibular component of the sciatic nerve, which supplies part of the hamstrings as well as the tibialis anterior muscle.[61]

Clinical Pearls: Quadriceps

- Vital to proper gait

- Imperative for running

- Crucial when going up and down stairs

- The rectus femoris is the only part of the quadriceps that crosses the knee and helps with hip flexion

- The full quadriceps muscle group will help with stability of the front of the thigh and hip joint

Clinical Pearls: Hamstrings

- Will help with walking

- The entire hamstring group will aid with knee flexion and thigh extension

- Imperative during daily and sporting activities

- Designed to help decelerate the limb during sports activities, walking, and running

- Restrain the tibia from moving in an anterior direction, thus sparing the ACL, and also help to maintain patellofemoral stability

References

1. Perry J. *Gait Analysis: Normal and Pathological Function.* Thorofare, NJ: SLACK Incorporated; 1992.

2. Lieb FJ, Perry J. Quadriceps function. An anatomical and mechanical study using amputated limbs. *J Bone Joint Surg Am.* 1968;50(8):1535-1548.

3. Raimondo RA, Ahmad CS, et al. Patellar stabilization: a quantitative evaluation of the vastus medialis obliquus muscle. *Orthopedics.* 1998;21(7):791-795.

4. Dutton M. *Orthopaedic Examination, Evaluation, and Intervention.* New York: McGraw-Hill; 2008.

5. Norkin CC, Levangie PK. *Joint Structure and Function: A Comprehensive Anaylsis*. 2nd ed. Philadelphia: FA Davis Co; 1992.

6. Besier TF, Fredericson M, et al. Knee muscle forces during walking and running in patellofemoral pain patients and pain-free controls. *J Biomech*. 2009;42(7):898-905.

7. Sakai N, Luo ZP, et al. In vitro study of patellar position during sitting, standing from squatting, and the stance phase of walking. *Am J Knee Surg*. 1996;9(4):161-166.

8. Sauvage LR Jr, Myklebust BM, et al. A clinical trial of strengthening and aerobic exercise to improve gait and balance in elderly male nursing home residents. *Am J Phys Med Rehabil*. 1992;71(6):333-342.

9. Mann RA, Hagy J. Biomechanics of walking, running, and sprinting. *Am J Sports Med*. 1980;8(5):345-350.

10. Alfredson H, Hedberg G, et al. High thigh muscle strength but not bone mass in young horseback-riding females. *Calcif Tissue Int*. 1998;62(6):497-501.

11. Escamilla RF, Francisco AC, et al. An electromyographic analysis of sumo and conventional style deadlifts. *Med Sci Sports Exerc*. 2002;34(4):682-688.

12. Wenos DL, Amato HK. Weight cycling alters muscular strength and endurance, ratings of perceived exertion, and total body water in college wrestlers. *Percept Mot Skills*. 1998;87(3 Pt 1):975-978.

13. Bloch RM. Figure skating injuries. *Phys Med Rehabil Clin N Am*. 1999;10(1):177-188, viii.

14. Kanehisa H, Nemoto I, et al. Strength capabilities of knee extensor muscles in junior speed skaters. *J Sports Med Phys Fitness*. 2001;41(1):46-53.

15. Cronin JB, Hansen KT. Strength and power predictors of sports speed. *J Strength Cond Res*. 2005;19(2):349-357.

16. Clemente CD. *Gray's Anatomy*. 30th ed. Philadelphia: Lea and Febiger; 1985.

17. Lyons K, Perry J, et al. Timing and relative intensity of hip extensor and abductor muscle action during level and stair ambulation. An EMG study. *Phys Ther*. 1983;63(10):1597-1605.

18. Tirosh O, Sparrow WA. Age and walking speed effects on muscle recruitment in gait termination. *Gait Posture*. 2005;21(3):279-288.

19. Ericson MO, Nisell R, et al. Quantified electromyography of lower-limb muscles during level walking. *Scand J Rehabil Med*. 1986;18(4):159-163.

20. Arnold AS, Komattu AV, et al. Internal rotation gait: a compensatory mechanism to restore abduction capacity decreased by bone deformity. *Dev Med Child Neurol*. 1997;39(1):40-44.

21. Bryant AL, Creaby MW, et al. Dynamic restraint capacity of the hamstring muscles has important functional implications after anterior cruciate ligament injury and anterior cruciate ligament reconstruction. *Arch Phys Med Rehabil*. 2008(12):2324-2331.

22. Rudolph KS, Axe MJ, et al. Dynamic stability in the anterior cruciate ligament deficient knee. *Knee Surg Sports Traumatol Arthrosc*. 2001;9(2):62-71.

23. Rudolph KS, Snyder-Mackler L. Effect of dynamic stability on a step task in ACL deficient individuals. *J Electromyogr Kinesiol*. 2004;14(5):565-575.

24. Stokes IA, Abery JM. Influence of the hamstring muscles on lumbar spine curvature in sitting. *Spine (Phila Pa 1976)*. 1980;5(6):525-528.

25. Dvir Z. Isokinetics: *Muscle Testing, Interpretation and Clinical Application*. 2nd ed. Philadelphia: Churchille Livingstone; 2004.

26. Stafford MG, Grana WA. Hamstring/quadriceps ratios in college football players: a high velocity evaluation. *Am J Sports Med*. 1984;12(3):209-211.

27. Larrat E, Kemoun G, et al. Isokinetic profile of knee flexors and extensors in a population of rugby players. *Ann Readapt Med Phys.* 2007;50(5):280-286.

28. Fry AC, Powell DR. Hamstring/quadricep parity with three different weight training methods. *J Sports Med Phys Fitness.* 1987;27(3):362-367.

29. Prietto CA, Caiozzo VJ. The in vivo force-velocity relationship of the knee flexors and extensors. *Am J Sports Med.* 1989;17(5):607-611.

30. Ivy J, Withers R, Brose G, Maxwell B, Costill D. Isokinetic contractile properties of the quadriceps with relation to fiber type. *Eur J Appl Physiol.* 1981;47(3):247-255.

31. Hiemstra LA, Webber S, et al. Hamstring and quadriceps strength balance in normal and hamstring anterior cruciate ligament-reconstructed subjects. *Clin J Sport Med.* 2004;14(5):274-280.

32. Hollinshead WH, Rosse C. *Textbook of Anatomy.* 4th ed. Philadelphia: Harper & Row; 1985.

33. Warfel J. *The Extremities: Muscles and Motor Points.* Philadelphia: Lea and Febiger; 1985.

34. Standring S. *Gray's Anatomy: The Anatomical Basis of Clinical Practice.* 40th ed. New York: Churchill Livingstone; 2008.

35. DeLee JC, Drez D Jr. *Orthopaedic Sports Medicine. Principles and Practice.* Philadelphia: WB Saunders Co; 1994.

36. Hurley MV, Jones DW, et al. Arthrogenic quadriceps inhibition and rehabilitation of patients with extensive traumatic knee injuries. *Clin Sci* (Lond). 1994;86(3):305-310.

37. Tsuno MM, Shu GJ. Myositis ossificans. *J Manipulative Physiol Ther.* 1990;13(6):340-342.

38. Haim A, Yaniv M, et al. Patellofemoral pain syndrome: validity of clinical and radiological features. *Clin Orthop Relat Res.* 2006;451:223-228.

39. Mason JJ, Leszko F, et al. Patellofemoral joint forces. *J Biomech.* 2008;41(11):2337-2348.

40. Mohr KJ, Kvitne RS, et al. Electromyography of the quadriceps in patellofemoral pain with patellar subluxation. *Clin Orthop Relat Res.* 2008;(415):261-271.

41. Henry JH Conservative treatment of patellofemoral subluxation. *Clin Sports Med.* 1989;8(2):261-278.

42. Fulkerson JP. Diagnosis and treatment of patients with patellofemoral pain. *Am J Sports Med.* 2002;30(3):447-456.

43. Guo K, Ye Q, et al. Selective training of the vastus medialis muscle using electrical stimulator for chondromalacia patella. *Zhongguo Yi Xue Ke Xue Yuan Xue Bao.* 1996;18(2):156-160.

44. Konrath GA, Chen D, et al. Outcomes following repair of quadriceps tendon ruptures. *J Orthop Trauma.* 1998;12(4):273-279.

45. Elias JJ, Cech JA, et al. Reducing the lateral force acting on the patella does not consistently decrease patellofemoral pressures. *Am J Sports Med.* 2004;32(5):1202-1208.

46. Lempainen L, Sarimo J, et al. Proximal hamstring tendinopathy: results of surgical management and histopathologic findings. *Am J Sports Med.* 2009;37(4):727-734.

47. Lysholm J, Wiklander J. Injuries in runners. *Am J Sports Med.* 1987;15(2):168-171.

48. McGregor C, Ghosh S, et al. Traumatic and overuse injuries of the ischial origin of the hamstrings. *Disabil Rehabil.* 2008;30(20-22):1597-1601.

49. Orchard J. Management of muscle and tendon injuries in footballers. *Aust Fam Physician.* 2003;32(7):489-493.

50. Agre JC. Hamstring injuries. Proposed aetiological factors, prevention, and treatment. *Sports Med.* 1985;2(1):21-33.

51. Melnyk M, Gollhofer A. Submaximal fatigue of the hamstrings impairs specific reflex components and knee stability. *Knee Surg Sports Traumatol Arthrosc.* 2007;15(5):525-532.

52. Tsepis E, Vagenas G, et al. Hamstring weakness as an indicator of poor knee function in ACL-deficient patients. *Knee Surg Sports Traumatol Arthrosc.* 2004;12(1):22-29.

53. Simonsen EB, Magnusson SP, et al. Can the hamstring muscles protect the anterior cruciate ligament during a side-cutting maneuver? *Scand J Med Sci Sports.* 2000;2:78-84.

54. Sarimo J, Lempainen L, et al. Complete proximal hamstring avulsions: a series of 41 patients with operative treatment. *Am J Sports Med.* 2008;36(6):1110-1115.

55. Schache AG, Koulouris G, et al. Rupture of the conjoint tendon at the proximal musculotendinous junction of the biceps femoris long head: a case report. *Knee Surg Sports Traumatol Arthrosc.* 2008;16(8):797-802.

56. Folsom GJ, Larson CM. Surgical treatment of acute versus chronic complete proximal hamstring ruptures: results of a new allograft technique for chronic reconstructions. *Am J Sports Med.* 2008;36(1):104-109.

57. Cohen S, Bradley J. Acute proximal hamstring rupture. *J Am Acad Orthop Surg.* 2007;15(6):350-355.

58. Kwong Y, Patel J, et al.Spontaneous complete hamstring avulsion causing posterior thigh compartment syndrome. *Br J Sports Med.* 2006;40(8):723-724; discussion 724.

59. Schilders E, Bismil Q, et al. Partial rupture of the distal semitendinosus tendon treated by tenotomy: a previously undescribed entity. *Knee.* 2006;13(1):45-47.

60. Alioto RJ, Browne JE, et al. Complete rupture of the distal semimembranosus complex in a professional athlete. *Clin Orthop Relat Res.* 1997;(336):162-165.

61. Hernesman SC, Hoch AZ, et al. Foot drop in a marathon runner from chronic complete hamstring tear. *Clin J Sport Med.* 2003;13(6):365-368.

10

THE GASTROCNEMIUS/SOLEUS AND FIBULARIS LONGUS AND BREVIS

On your mark, get set, go! The runner takes off down the lane as fast as she can go to reach the finish line, and hopefully have a successful outcome. In order for this to happen, the runner needs to coordinate a number of muscles to propel out of the blocks, 2 of which are the gastrocnemius and the soleus. These muscles, known collectively as the triceps surae muscle group, make up the vast bulk of the calf muscles. The triceps surae collectively is a critical group of muscles for a variety of functional activities. When you walk up stairs, the triceps surae pushes you up to the next step and during walking, the calf muscle group is a necessity to maintain normal velocity and normal step and stride length.[1] While the triceps surae muscle group is propelling you through space, the fibularis longus and brevis are working to keep you upright. The laterally placed fibularis longus and brevis (formerly known as the peroneus longus and brevis) have an important role in gait and balance. In this chapter, we will explore both muscle groups and explain their many needed functions.

Function and Functional Anatomy

The gastrocnemius is a strong muscle with 2 heads of origin on the femur and thus crosses both the knee and ankle joints (Figures 10-1A and B). The soleus is a large, more deeply placed muscle of the posterior calf that crosses only the ankle. Flexing the knee will relax the gastrocnemius at the ankle, but not the soleus. Both muscles are active during walking, running, and stair climbing. During running, the medial head of gastrocnemius will contribute more than the lateral head (Figures 10-2B and C).[2,3]

One of the most important functions that the gastrocnemius and the soleus perform is controlling the forward momentum of the body during the gait cycle when they act in a closed kinetic chain.[1,4] This muscle group has been reported to be important for balance and position sense, pelvic control,[5-7] and during sporting activities.[8] As discussed earlier, the gastrocnemius/soleus group is very active during the walking cycle.[1] In walking, the gastrocnemius/soleus complex contracting concentrically generates about 93% of the push-off torque needed to continue stepping forward smoothly.[1] Also, during the gait sequence, the plantar flexors (gastrocnemius/soleus) are active from mid-stance to terminal stance in order to eccentrically limit

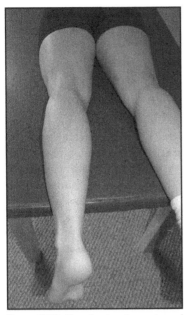

Figure 10-1B. Schematic of the left gastrocnemius with arrows showing the forces from the insertion at the ankle to the medial and lateral muscle heads. (Reprinted with permission from Primal Pictures, 2009.)

Figure 10-1A. The relaxed gastrocnemius and soleus underneath in prone.

Figure 10-2A. The right gastrocnemius/soleus muscle contracted in prone.

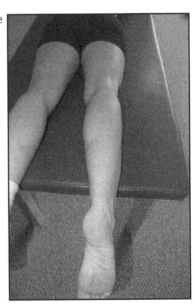

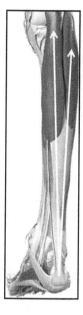

Figure 10-2B and C. (B) The soleus muscle and (C) gastrocnemius (both heads, right side). (Reprinted with permission from Primal Pictures, 2009.)

ankle dorsiflexion and knee flexion.[1,9,10] The medial gastrocnemius (largest head) will fire first and the lateral head will contract later in mid-stance, while the soleus first begins its action at the end of the loading response (or foot flat) and continues through mid-stance and through terminal stance with its largest contribution occurring in terminal stance (see Figures 10-2A, B, and C).[1]

Beside walking and other gravity-resisting functions, like going up and down stairs, the gastrocnemius/soleus will be also active during running.[11-13] When a person stands in an upright position, there is a close coordinated effort between activation of triceps surae and the hamstrings. The 2-joint gastrocnemius allows power generated at the hip and knee to be transferred to the ankle, and vice versa.[14] During the running cycle, contraction of the triceps surae will propel the foot downward to push off when each stride occurs, maintaining the forward momentum of the body's mass through the running cycle.[15]

The fibularis longus and brevis function (Figures 10-3A and B) as primary evertors and abductors of the foot[16] and also will serve as secondary plantar flexors (Figures 10-4A and B).[17] Both muscles are needed for balance and also during the gait cycle.[1] Fibularis longus can stabilize

Figure 10-3A. The fibularis longus and brevis (under the longus) relaxed.

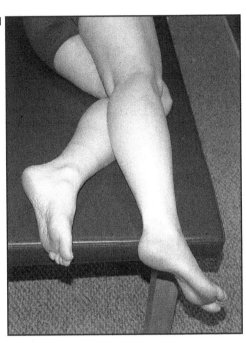

Figure 10-3B. The fibularis longus and brevis contracted. Fibularis muscle shown as noted by the black dot on the longus origination site at the fibular head.

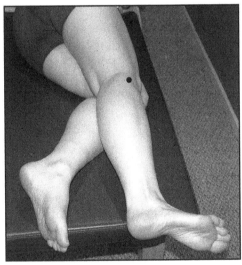

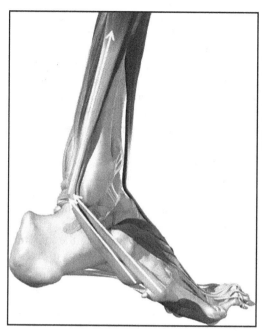

Figure 10-4A. The fibular longus and brevis tendons. The far lateral attached tendon to the fifth metatarsal is the fibularis brevis and the more medial tendon running underneath the foot is the fibularis longus. The arrows show the forces going from the insertion to the origin. (Reprinted with permission from Primal Pictures, 2009.)

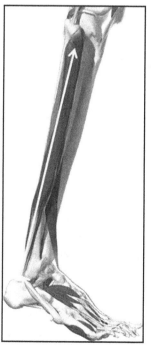

Figure 10-4B. Full length of the muscle and the fibularis longus shown with the forces from the origin to the insertion of the muscle (first arrow for the fibularis brevis and the top arrow for the fibular longus to the origin at the fibular head). (Reprinted with permission from Primal Pictures, 2009.)

the medial longitudinal arch of the foot and add to stability at the first metacarpophalangeal joint, through its distal attachment.[18] The fibularis longus and brevis will aid in balancing the foot during daily activities and also work in concert with the triceps surae during the gait cycle.[1,19] Specifically, both muscles are recruited during loading of the forefoot of the gait cycle, with the fibularis longus being activated first, followed by the fibularis brevis.[1,20,21] As the gait cycle progresses, the fibularis longus and brevis are less active and will relax somewhat during pre-swing to mid pre-swing (see Figure 10-4A).[1] During the gait cycle, the fibularis muscles will be recruited to a variable extent, depending on how a person walks, to stabilize the foot and prevent too much inversion at the foot and ankle, as well as to provide a lateral pull on the leg to counteract the effect of the gravity vector in single limb support, thus aiding balance.[16] The fibularis muscles are more active during sports, whether during running activities or during a skating routine, the muscles' role in controlling the foot and providing balance becomes even more important.[22]

Origin and Insertion

Origin of the Gastrocnemius and Soleus and the Fibularis Longus and Brevis Muscles	Insertion of the Gastrocnemius and Soleus and Fibularis Longus and Brevis Muscles
GASTROCNEMIUS: The gastrocnemius muscle occupies the most superficial division of the posterior fascial compartment of the leg. Gastrocnemius is a large muscle with 2 heads of origin. The larger of the 2 heads is the medial.[2,16] Both heads arise from the posterior condyles of the femur by large flat tendons (see Figure 10-1B).[3,16]	The medial and lateral heads of the gastrocnemius unite before joining the tendon of the soleus and then converge together to insert into the posterior aspect of the calcaneus.[16,23,24]
SOLEUS: The soleus lies just deep to the gastrocnemius. It is flatter than and has a larger volume of muscle compared to the gastrocnemius. It takes origin from the posterior aspect of the fibular head, the medial portion of the tibia, and the soleal line on the posterior tibia (see Figures 10-2B and C).	The soleus muscle courses caudally to join the gastrocnemius muscle fibers to insert on the posterior aspect of the calcaneus.[3,16]
FIBULARIS LONGUS: This muscle is located on the lateral side of the leg, in the proximal position, and will originate from upper two-thirds and the head of the fibula.[2,16] Between the anterior and posterior leg muscles, the fibularis longus will have an origination along with some of its fibers coming off from the deep fascia of the upper portion of the tibia (see Figures 10-3A and 10-4A).[2,23]	The fibularis longus muscle travels downward and via a long tendonous attachment, courses along the lateral aspect of the lateral malleolus. The tendon then turns obliquely across the calcaneus to cross the lateral aspect of the cuboid bone. The fibularius longus tendon curves under and around the cuboid bone, moving medially to the sole of the foot to finally attach itself to the base of the first metatarsal bone and first cuneiform bone (see Figure 10-4A).[2,16]

(Continued)

Origin of the Gastrocnemius and Soleus and the Fibularis Longus and Brevis Muscles	Insertion of the Gastrocnemius and Soleus and Fibularis Longus and Brevis Muscles
FIBULARIS BREVIS: The fibularis brevis is the deeper of the 2 lateral fibular muscles and will arise under the fibularis longus from the distal two-thirds of the fibular body and also from the intermuscular septum (see Figures 10-3B and 10-4B).[2,16]	The fibularis brevis fibers move vertically downward to the lateral malleolus, forming a tendon that also courses around the lateral malleolus just anterior to the tendon of the fibularis longus (see Figure 10-4B).[16] The fibularis brevis tendon turns forward on the lateral aspect of the calcaneus to attach to the base of the fifth metatarsal bone.[2]
ACTION OF THE GASTROCNEMIUS AND SOLEUS MUSCLES AND THE FIBULARIS LONGUS AND BREVIS	
The gastrocnemius and the soleus is a group of muscles that is considered plantarflexors, which will help to push the foot downward.[3,23]	
The fibularis longus and brevis will act as evertors of the foot and have a small secondary action of plantar flexion. They aid in balance and will help with maintaining the longitudinal and transverse arches of the foot.[3,16,17]	

Pathomechanics of Common Injuries

One of the most common injuries to the conjoint tendon of gastrocnemius/soleus, the Achilles tendon, is tendinitis.[25,26] Achilles tendon problems can occur during sports activities[27-29] or from a common disease such as diabetes, or simply with aging.[30] An area of the tendon about 2 to 6 cm above the insertion on the calcaneus is an area that is not highly vascularized.[3] Because this tendon is subjected to heavy loads from both the gastrocnemius and soleus, the tendon may become overloaded and inflamed.[31]

The full catalog of common calf injuries ranges from **simple tendinitis to spontaneous rupture**.[3,31,32] Injuries that occur due to sports or recreational activities can occur to the conjoint tendon from a push-off in running, cutting, or jumping. In all cases, the tendon is more subject to injury due to a ballistic movement that results in sudden high loading on the tendon.[33] In other cases, with aging, the tendon **degenerates (tendinosis)**, is easily injured, and does not heal. With each stressor, it continues to become injured, which can lead to the highly organized collagen of the tendon being replaced by much weaker scar tissue. This situation may predispose a person to Achilles tendon rupture.[34-37]

Haglund-Akerlind et al[38] showed that runners frequently have Achilles trouble and the more they run, the more severely they can injure the tendon. As the process of **degeneration** occurs for this tendon, there is a decrease in concentric and eccentric torque output as well as the active and passive range of motion (the muscle's functional excursion) for the gastrocnemius/soleus complex.[38] Therefore, the biomechanics of the leg and foot are affected due to the smaller functional excursion and the reduced eccentric and concentric control provided by the gastrocnemius/soleus group.[39] Lastly, another common injury to the Achilles tendon, just like any other tendon, is **calcification,** which leads the tendon to a place where it is less pliable and more prone to injury.[40]

When injuries to the tendon occur, recent studies have indicated that one of the best ways to view the muscle tendon junction is ultrasonic imaging.[41] Magnetic resonance imaging has been the gold standard for many years but ultrasound is less expensive and will show muscle tears and tendon injuries.[41,42]

The fibularis longus and brevis are subject to certain injuries as well. Because they are located on the outside of the leg and directly overlie a bone, these muscles are subject to injuries involving direct trauma as well as internal pathologies. Some of these injuries may require bracing, taping, or even surgery. The most common fibularis injury is **tendinitis.**[43] Tendinitis is likely to be seen in the acute stages due to pain and commonly occurs with overuse, as during running, or sporting events, but can be the result of a single traumatic event such as tripping or falling off a curb or step. In some cases, an injury diagnosed as a tendinitis may in reality be a **partial tendon rupture.**[44,45] These are not uncommon especially for sporting individuals, such as basketball or football players, and usually require more stabilization (taping, bracing) than simple tendinitis. Internal physiologic changes in the tendons, as occur with aging, can predispose people to more frequent injury, which may present as a chronic tendinitis but is more properly called **tendinosis,** a condition that involves degeneration and scarring in the tendon sheath.[44] A rare but significant condition that can occur is one of **tenosynovitis** of the fibularis longus tendon as it courses along its path to insert on the first metatarsal. Swelling and inflammation of the tendon sheath causes the tendon to be pinched, causing pain as it moves around the cuboid bone through a bony tunnel. As pain ensues, function is compromised.[46]

Clinical Pearls: Gastrocnemius and Soleus

- Active during walking, running, and jumping

- Needed during walking up stairs and for most functional activities

- Helps to support the ankle and knee joints

- Needed for proper lower extremity biomechanics

Clinical Pearls: Fibularis Longus and Brevis

- Aid in walking, running, and jumping by providing balance and maintaining the longitudinal and transverse arches of the foot

- Stabilize the forefoot during push-off in the gait cycle

- Maintain a healthy foot balance for inversion and eversion during daily activities

References

1. Perry J. *Gait Analysis: Normal and Pathological Function.* Thorofare, NJ: SLACK Incorporated; 1992.

2. Hollinshead WH, Rosse C. *Textbook of Anatomy.* 4th ed. Philadelphia: Harper & Row; 1985.

3. Dutton M. *Orthopaedic Examination, Evaluation, and Intervention.* New York: McGraw-Hill; 2008.

4. Petersen N, Morita H, et al. Modulation of reciprocal inhibition between ankle extensors and flexors during walking in man. *J Physiol.* 1999;520(Pt 2):605-619.

5. Danna-Dos-Santos A, Shapkova EY, et al. Postural control during upper body locomotor-like movements: similar synergies based on dissimilar muscle modes. *Exp Brain Res.* 2009;193(4):565-579.

6. Chen HL, Lin YC, et al. The effect of ankle position on pelvic floor muscle contraction activity in women. *J Urol.* 2009;181(3):1217-1223.

7. Ciccarelli O, Toosy AT, et al. Identifying brain regions for integrative sensorimotor processing with ankle movements. *Exp Brain Res.* 2005;166(1):31-42.

8. Wilk BR, Fisher KL, et al. Defective running shoes as a contributing factor in plantar fasciitis in a triathlete. *J Orthop Sports Phys Ther.* 2000;30(1):21-28; discussion 29-31.

9. Ericson MO, Nisell R, et al. Quantified electromyography of lower-limb muscles during level walking. *Scand J Rehabil Med.* 1986;18(4):159-163.

10. Lewis CL, Ferris DP. Walking with increased ankle pushoff decreases hip muscle moments. *J Biomech.* 2008;41(10):2082-2089.

11. Chumanov ES, Wall-Scheffler C, et al. Gender differences in walking and running on level and inclined surfaces. *Clin Biomech (Bristol, Avon).* 2008;23(10):1260-1268.

12. McGillivray DG, Garland T Jr, et al. Changes in efficiency and myosin expression in the small-muscle phenotype of mice selectively bred for high voluntary running activity. *J Exp Biol.* 2009;212(Pt 7):977-985.

13. Neptune RR, Sasaki K. Ankle plantar flexor force production is an important determinant of the preferred walk-to-run transition speed. *J Exp Biol.* 2005;208(Pt 5):799-808.

14. van Soest AJ, Schwab AL, et al. The influence of the biarticularity of the gastrocnemius muscle on vertical-jumping achievement. *J Biomech.* 1993;26(1):1-8.

15. Ounpuu S. The biomechanics of running: a kinematic and kinetic analysis. *Instr Course Lect.* 1990;39:305-318.

16. Clemente CD. *Gray's Anatomy.* 30th ed. Philadelphia: Lea and Febiger; 1985.

17. Norkin CC, Levangie PK. *Joint Structure & Function: A Comprehensive Analysis.* 2nd ed. Philadelphia: FA Davis Co; 1992.

18. Johnson C, Christensen J. Biomechanics of the first ray part I. The effects of peroneus longus function: A three-dimensional kinematic study on a cadaver model. *J Foot Ankle Surg.* 1999;38:313-321.

19. McLoda TA, Hansen AJ. Effects of a task failure exercise on the peroneus longus and brevis during perturbed gait. *Electromyogr Clin Neurophysiol.* 2005;45(1):53-58.

20. Title CI, Jung HG. The peroneal groove deepening procedure: a biomechanical study of pressure reduction. *Foot Ankle Int.* 2005;26(6):442-448.

21. Houtz SJ, Walsh FP. Electromyographic analysis of the functions of the muscles acting on the ankle during weightbearing with special reference to the triceps surae. *J Bone Joint Surg Am.* 1959;41-A:1469-1481.

22. Wind WM, Rohrbacher BJ. Peroneus longus and brevis rupture in a collegiate athlete. *Foot Ankle Int.* 2001;22(2):140-143.

23. Warfel J. *The Extremities: Muscles and Motor Points.* Philadelphia: Lea and Febiger; 1985.

24. Standring S. *Gray's Anatomy: The Anatomical Basis of Clinical Practice.* 40th ed. New York: Churchill Livingstone; 2008.

25. Longo UG, Ronga M, et al. Achilles tendinopathy. *Sports Med Arthrosc.* 2009;17(2):112-126.

26. Christenson RE. Effectiveness of specific soft tissue mobilizations for the management of Achilles tendinosis: single case study—experimental design. *Man Ther.* 2007;12(1):63-71.

27. Oden RR. Tendon injuries about the ankle resulting from skiing. *Clin Orthop Relat Res.* 1987;(216):63-69.

28. McCrory JL, Martin DF, et al. Etiologic factors associated with Achilles tendinitis in runners. *Med Sci Sports Exerc.* 1999;31(10):1374-1381.

29. Segesser B, Goesele A, et al. The Achilles tendon in sports. *Orthopade.* 1995;24(3):252-267.

30. Batista F, Nery C, et al. Achilles tendinopathy in diabetes mellitus. *Foot Ankle Int.* 2008;29(5):498-501.

31. Reynolds RD, Nelson LB, et al. Large refractive change after strabismus surgery. *Am J Ophthalmol.* 1991;111(3): 371-372.

32. Bryan Dixon J. Gastrocnemius vs. soleus strain: how to differentiate and deal with calf muscle injuries. *Curr Rev Musculoskelet Med.* 2009;2(2):74-77.

33. Smerdelj M, Madjarevic M, et al. Overuse injury syndromes of the calf and foot. *Arh Hig Rada Toksikol.* 2001;52(4):451-464.

34. Amlang MH, Zwipp H. Damage to large tendons: Achilles, patellar and quadriceps tendons. *Chirurg.* 2006;77(7):637-649, quiz 649.

35. Mulvaney S. Calf muscle therapy for Achilles tendinosis. *Am Fam Physician.* 2003;67(5):939; author reply 939-940.

36. Fahlstrom M, Jonsson P, et al. Chronic Achilles tendon pain treated with eccentric calf-muscle training. *Knee Surg Sports Traumatol Arthrosc.* 2003;11(5):327-333.

37. Hart LE. Exercise and soft tissue injury. *Baillieres Clin Rheumatol.* 1994;8(1):137-148.

38. Haglund-Akerlind Y, Eriksson E. Range of motion, muscle torque and training habits in runners with and without Achilles tendon problems. *Knee Surg Sports Traumatol Arthrosc.* 1993;1(3-4):195-199.

39. Silbernagel KG, Gustavsson A, et al. Evaluation of lower leg function in patients with Achilles tendinopathy. *Knee Surg Sports Traumatol Arthrosc.* 2006;14(11):1207-1217.

40. Mohr W, Hersener J, et al. Suture granuloma with calcium pyrophosphate deposits. *Z Orthop Ihre Grenzgeb.* 1990;128(2):134-138.

41. Bianchi S, Poletti PA, et al. Ultrasound appearance of tendon tears. Part 2: lower extremity and myotendinous tears. *Skeletal Radiol.* 2006;35(2):63-77.
42. Whittaker RG, Ferenczi E, et al. Myotonic dystrophy: practical issues relating to assessment of strength. *J Neurol Neurosurg Psychiatry.* 2006;77(11):1282-1283.
43. Sammarco GJ. Peroneal tendon injuries. *Orthop Clin North Am.* 1994;25(1):135-145.
44. Yao L, Tong DJ, et al. MR findings in peroneal tendonopathy. *J Comput Assist Tomogr.* 1995;19(3):460-464.
45. Slater HK. Acute peroneal tendon tears. *Foot Ankle Clin.* 2007;2(4):659-674, vii.
46. Bruce WD, Christofersen MR, et al. Stenosing tenosynovitis and impingement of the peroneal tendons associated with hypertrophy of the peroneal tubercle. *Foot Ankle Int.* 1999;20(7):464-467.

SECTION III

The Spine

Benjamin S, Bechtel RH, Conroy VM.
Cram Session in Functional Anatomy:
A Handbook for Students & Clinicians (pp. 87-110)
© 2011 Taylor & Francis Group.

11

REVIEW OF SPINAL ANATOMY

Research on lumbar forces and loads during functional activities is in its infancy, and most models of lumbar function are based on incomplete and/or highly circumscribed data sets. It is necessary, therefore, to rely on clinical experience to fill in the gaps in our present understanding of lumbar biomechanics. This chapter will discuss the anatomy of the spine and its components with consideration of the current understanding of their roles in function and dysfunction.

The spine is a flexible mechanical system that has several important jobs to perform. It must protect the spinal cord and nerves and also allow us to move in many directions. The spine must bear weight allowing us to stand upright, as well as having us bend and twist so that we can perform functional activities in our environment. For motions to occur, the bones of the spine (the vertebrae) must be separated by a flexible connector, which is called the intervertebral disc. Because significant weightbearing occurs through the vertebral bodies, the intervertebral disc must be not only flexible, but also tough and tear resistant. All of the intervertebral joints have discs with the exception of the occiput on C1 and C1 on C2. In the adult, the discs account for about 25% of a person's height.[1]

Function of the Spine

Functionally, the vertebra is divided into anterior and posterior elements. The anterior element is the body of the vertebra and the moveable posterior elements are the facets that guide and limit motion. There are a total of 33 vertebra in the spine.[2,3] The spine as a mechanical system is a flexible, segmented rod. In humans, the spine is for support and gives humans the opportunity to either flee from danger or defend themselves. The spine must protect the spinal cord and the nerves that control the muscles that allow us to move about as well as help us bend, twist, and allow us to adapt rapidly to the environment in which we find ourselves.

While performing these tasks, the upright spine is exposed to significant forces and moments applied by the extremities, as well as by the spinal muscles and gravity. The upper limb girdle attaches to the spine at the sternoclavicular joint anteriorly (attached to the axial skeleton) and by muscular connections to the ribs and vertebral column posteriorly. The lower limb girdle attaches through the pelvis, containing the pubic symphysis and sacroiliac joints, with one point of contact with the spine, at the L5-S1 intervertebral joint. The joints here, therefore, are subjected to more stress than most other spinal joints. The disc between L5 and S1 is a frequent source of problems.[4] The concept of spine stability is dynamic. Panjabi[5] has envisioned a 3-part system that is responsible for stabilizing the spinal joints against the forces that are applied to them. The system consists of passive elements like ligaments and joint capsules, active elements like the muscles, and a central controller mechanism shared between the central and peripheral nervous systems. The concept of the "neutral zone" is crucial to understanding how the system works. The neutral zone is the zone of motion that is available at each spinal segment before any resistance is felt. It is entirely under the control of muscles.

The spine has 2 major ligaments that run along the anterior and posterior elements, which are referred to as the anterior and posterior longitudinal ligaments. The 2 ligaments will limit extension and flexion, respectively. The posterior longitudinal ligament lies directly behind the intervertebral disc and directly anterior to the spinal cord or nerve roots in the neural canal.[6] In the lumbar region, this ligament is less well developed, which has been related to the increased frequency of posterior disc herniations in this part of the spine.[7,8] Between adjacent lamina, on either side of the neural canal, are the paired ligamenta flava. These ligaments lie just anterior to the facet joints and form the posterior wall of the intervertebral foramen through which the spinal nerves exit.

Muscles of the Spine

Movements and muscle activity will apply compressive and shear loads on the spine[5,9] with the highest compressive forces found in the lumbar spine (Figures 11-1A and B).[10] Over time, the forces applied to the spine can break down its components and cause pain as well as observable changes in motion.[11] We will now look at the muscles of the lumbar spine and also their function.

One immediately salient point that confronts anyone who has ever dissected a human cadaver is the complex web of mechanical connections between different tissues, layers, and systems within the body. For example, dissect almost any fascial layer and you will find muscle fibers attaching along its length. **Fascia is not so much a sheath around muscles as a conduit for the conveyance of muscle force within the body.** Although we often visualize a muscle as having discrete points of origin and insertion, and most current biomechanical models subscribe to this simplification, the distribution of muscle forces in the body is more complex because of

Figure 11-1A. Relaxed lower back muscles (multifidus).

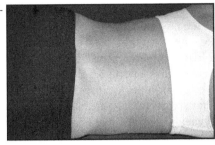

Figure 11-1B. The lumbar multifidus (right) muscle schematic with the arrows showing the muscle forces. (Reprinted with permission from Primal Pictures, 2009.)

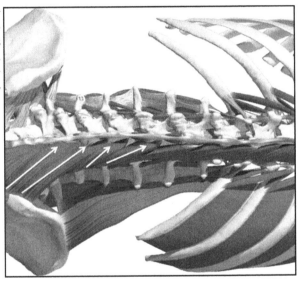

the multiple interfaces between muscle, fascia, and surrounding structures. This is most evident in the muscles of the spine.

Although the origin and insertion simplification is taught in most medical and allied health schools, we need to realize that in the body, muscle force is applied over a broader area than would be immediately obvious by considering only origin and insertion. The significant effect of mechanical disbursement of muscle force through fascial attachments has recently been demonstrated in the rat model.[12] One implication of this realization is that proprioceptive inputs to the central nervous system associated with muscle contraction do not arise only from tissues directly in the path from origin to insertion but may arise in tissues outside this path as well; thus changes in muscle activity have implications for tissues and systems in a broad area surrounding the muscle's anatomic path. Conversely, the receptive field of muscle proprioceptors is probably a good deal larger than we currently imagine.

Spine Muscle Organization

Muscles of the spine can be organized anatomically, by virtue of their connections to fascial planes, into superficial, intermediate, and deep groups (see Figures 11-1A and B). The superficial muscles play a major role in movement of the extremities. This group includes the trapezius, latissimus dorsi, rhomboids, and levator scapulae. The intermediate group plays a role in controlling rib motion. The intermediate group consists of 2 muscles, the serratus posterior superior and the serratus posterior inferior, which lie deep to the rhomboids and latissimus dorsi, respectively. The deep group is what many people think of as the "back muscles." These muscles are separated from the more superficial layers by the deeply placed thoracolumbar fascia. The role of the thoracolumbar fascia in function and dysfunction of the back muscles cannot be overemphasized. We will discuss it again in the following sections. The deep group of back muscles consists of the erector spinae, generally long muscles oriented longitudinally along the spine, and therefore also called the paraspinals, as well as the transversospinales, the interspinales, and intertransversarii (Figures 11-1B and 11-2B). The erector spine group is further divided from the lateral to its medial orientation as the iliocostales, the longissimus medially, and the spinales centrally (Figures 11-1A and 11-2A). All of the muscles in the erector spinae group function to straighten or extend the trunk.[2,3,13,14]

The transversospinales muscles consist of 3 muscle groups that lie deep to the erector spinae. These muscles run from transverse process to spinous process, across 1 or 2 motion segments (Figure 11-2B). The most superficial muscles of this group, the semispinalis muscles, are only found in the thoracic spine.[3,15] Other superficial muscles that are noted structural components are the semispinalis capitis and the cervicis. The next deepest muscles are the multifidus, which are easily visualized and the best developed in the lumbar region (see Figure 11-1B). The deepest muscles of the transversospinales group are the rotators. The short rotators span 1 motion segment, and the long rotators span 2 motion segments. Muscles in the transversospinales group can cause the spine to extend and side bend toward the active muscle and rotate away from the active muscle. All of the back muscles are innervated by dorsal branches of spinal nerves that emerge at adjacent segmental levels.[16,17] As these muscles function, a side effect of their actions is placed on the spine.

The longitudinally oriented lumbar muscle contraction will create spinal compression and clinicians need to design rehabilitation programs that do encourage spinal stabilization by muscle activity while minimizing compressive loading on the intervertebral discs.[18-20] For example, one

Figure 11-2A. The contracted lumbar multifidus in prone.

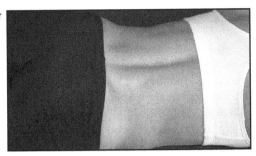

Figure 11-2B. The left multifidus schematic with arrows showing the muscle forces. (Reprinted with permission from Primal Pictures, 2009.)

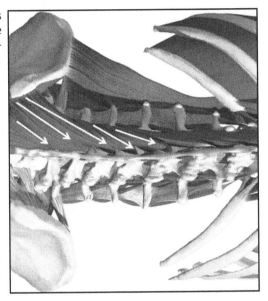

model showed that muscle response to increasing lifting load increases spine stability at a faster rate than the increase in vertebral compression force. However, this study did not take into account segmental mechanics, particularly the destabilizing effect of one or more segments with increased neutral zones as are typically seen in patients.[5]

Local and Global Muscles

Current research suggests that spine muscles function both to stabilize the spine and to produce movement.[21] Additionally, through their fascial attachments, both upper and lower extremity muscles can modulate spinal stability while producing movement. Some have suggested that spinal muscles should be divided functionally into **local** (primarily stabilizing) and **global** (primarily movement producing) muscles. According to Bergmark,[22] the local system consists of "All muscles which have their origin or insertion at the vertebrae, with the exception of the psoas...", whereas the global system consists of "muscles which transfer load directly between the thoracic cage and the pelvis."[22] From a clinical perspective, there is a good deal to be gained by understanding the dichotomy of local versus global. Local muscles tend to be deeper and generally lie closer to the axis of rotation of the motion segment while global muscles tend to be more superficial and further away from the axis. Local muscles and their associated fascia stiffen motion segments, while global muscles are suited to apply torque to the motion segments and potentially destabilize them. One significant exception to this neat categorization is the psoas major muscle, a prime hip flexor. The psoas takes origin from the anterior surfaces of the lumbar vertebrae bilaterally and inserts on the lesser trochanter of the femur, along with the iliacus. However, psoas contraction has the effect of a global muscle on the lumbar spine, increasing its susceptibility to injury.[23] A longer muscle, if unipennate like psoas and most of the longitudinal spine muscles, also tends to have more muscle fibers in series, giving it some unique properties such as the ability to better maintain force output at higher contraction velocities, compared to shorter or multipennate muscles. This makes the global muscles well suited for accommodating sudden changes in position and leaving the slower, more tonic contractions to the local muscles.[24,25]

The Neural Control System

Both proprioceptive and nociceptive inputs have a direct effect on the neural control system. For example, in patients with sacroiliac pain, muscle recruitment strategies are altered during a supine straight leg raise[26] and during hip flexion in standing.[25,27] Further, it has been shown that there are direct effects of spinal column structures on muscle recruitment. Indahl et al[16] showed that lumbar facet capsule stretch inhibits the ipsilateral multifidus in a porcine model while Solomonow et al[8] showed that electrical stimulation of the supraspinous ligament resulted in facilitation of multifidus muscles in humans. A good summary of these interactions is provided by Holm et al.[24] In general, it appears that local muscles tend to be inhibited in low back pain, while global muscles are abnormally facilitated and this leads to poor motion patterns and muscle recruitment.

In our daily lives, we routinely call upon the global muscles to move us through the environment. We expect that the spinal stresses induced by these high-powered, fast-acting muscles will be counteracted by a functioning subset of motor units in the generally smaller, slower-acting local muscles.[28] To maintain this important balance, the local muscles must not be inhibited.[16,29] Failure to identify factors that could inhibit local muscle function may result in overstressing the spine during rehabilitation, adding insult to injury. In practice, we must make every effort to identify and treat hypo- or hypermobile (unstable) segments in the patient's spine and any peripheral dysfunctions that could result in altered biomechanics or pain, prior to prescribing exercises.

Evaluating the segmental neutral zone can usually be done fairly effectively by posterior-anterior glides in prone, which reveal both the quality of resistance to movement and any increase in anterior travel.[30] The hypomobile segment can be identified as an aberration in an otherwise mobile spine. It has been proposed by Meadows[31] that segmental hypo- and hypermobility are complimentary states of the same underlying pathology: a failure of the myofascial system to adequately regulate movements within the spinal neutral zone. Both a hypo- and a hypermobile segment demonstrate alterations in the motor control program governing spinal stability. Both can result in unwanted stress transfers to adjacent segments. Finding and fixing these dysfunctions should be one of the first priorities of the rehabilitation professional who is treating patients with back pain, whether of idiopathic or postsurgical origin, because through their influence on the neural control mechanism, they can have a significant impact on the patient's quality of life and the longevity of any spinal implant.

Function of the Lumbar Multifidus

For the purposes of this section, we will focus our attention to the multifidus muscles of the lumbar spine. The multifidus is the largest and most medial of all the lower back stabilizing muscles (see Figures 11-2A and B).[3,14,32] The multifidus is a repeating segmental muscle that traverses from the spinous process of one segmental level, spiraling downward and laterally to attach to the mamillary body 1 or 2 segments below, and at L5, to the iliac crest and the sacrum (see Figures 11-1B, 11-2A, and 11-2B).[3,23] When an individual presents with lower back pain, his or her lumbar multifidus muscles are often compromised and their function is lost or reduced.[18,33]

Anatomical Relationship

The multifidus muscle is the most medial and deep stabilizer of all back muscles and will aid in stabilizing the spine, and its muscular lever arm is long enough to take compression off of the spinal segments and disc material (see Figure 11-1B).[32-34] From the top downward, the multifidus attaches to the spinous process of L1 through L5 and obliquely laterally and downward to its insertion site on the mamillary body on the vertebrae below. While typically thought of as a segmental muscle, research has shown that the multifidus may span from 1 to 3 segments and is divided into superficial and deep parts (see Figure 11-1B).[3,32,34] Thus, the lumbar motion is controlled by a deep and superficial portion of the multifidus muscle. It appears that deeper fibers are more related to stability, while the more superficial fibers have a bigger role in producing movement.[34] Innervation for this muscle is the medial branch of the dorsal ramus[3,32] from all lumbar levels.[35] The lumbar multifidus is active in extension of the lumbar spine and also axial rotation ipsilaterally and contralaterally.[3,32,36] It also works synergistically with the deep abdominal muscles to minimize flexion in the spinal segments, through their shared connections to the thoracolumbar fascia. Multifidus thus acts as a check and balance to the global trunk flexors, so that any one segment does not flex too much, which could lead to lumbar dysfunction.[19,37]

Origin and Insertion

Origin of the Lumbar Multifidus Muscle	Insertion of the Lumbar Multifidus Muscle
The short or deep portion of the multifidus will have short fascicles that will arise from a particular dorsal surface of the particular vertebral bodies' lamina (see Figure 11-1B).	The fibers from the deep or short portion of the lumbar multifidus will travel in an oblique fashion to the mamillary processes of the vertebra 1 or 2 segments below the site of origin (see Figure 11-1B).[32,38,39]
The more superficial multifidus fibers will travel from a caudal/lateral origin of the originating spinous process (see Figure 11-2B).	The superficial fibers will attach themselves to the mamillary process of the vertebra below or to the ilium or sacrum.[2,32,39]
ACTION OF THE LUMBAR MULTIFIDUS	
The lumbar multifidus has 2 related jobs to perform. The deeper fibers, considered the "local" portion of the muscle, cannot produce much torque and are more tonically active to control movement.[9,28] The superficial fibers of the lumbar multifidus are considered the working "global" portion of the muscle[33] and will help produce lumbar motions, primarily lumbar extension and ipsilateral rotation.	

Pathomechanics of Common Injuries

Lower back pain is the one of the most common disabilities that causes people to visit physicians' offices.[40] Most acute lower back pain episodes will resolve in 2 to 3 weeks from the onset[18,41] but it is recognized that 2% to 3% of those cases will advance into what we call **chronic lower back pain,** which can persist for months or years.[42]

What is very interesting here is that after a first occurrence of lower back pain, the likelihood of a second episode or reoccurrence of the pain is very high, between 60% and 85%, primarily as a result of altered muscle function in the spine and consequent decreased spine dynamic stability.[18] If you are a sports person and you have torn the anulus, and then 6 weeks later return to the same sport without having rehabilitated multifidus, the chances are higher that you will hurt the disc again. MacDonald et al[43] postulate that a person loses a degree of back muscle control in the first episode of lower back pain, but with recurrences the loss of control increases even further. MacDonald[43] also showed that even during the periods when the person is not experiencing pain, his or her deep lumbar multifidus muscles are not firing properly to stabilize the spine (see Figures 11-2A and B). This lack of stability in your spine is like a ticking time bomb waiting to give you a nasty surprise, so focusing on the multifidus muscles early in the rehabilitation process is a very good idea.

When lower back pain is experienced, a patient will show a very high prevalence of multifidus atrophy, and this can affect his or her activity level or even attempts to perform an activity.[44] Hides et al[45] also showed that not only was the multifidus not recruited correctly, but its size was diminished as evidenced by a reduction in the **cross-sectional area**. Ulrich et al[46] reported via in vivo studies that when the lumbar multifidus is not firing, the stiffness and stability of the lumbar spine is reduced. Therefore, the lumbar multifidus will appear to be inhibited in lower back pain as the result of a disconnection from the **neurological feedback loop**, with a resulting decrease in spinal segmental stability.[44,45]

Lee et al[47] considered the effect of various postures as they related to the cross-sectional area of the lumbar multifidus muscle. Patients with lower back pain exhibited decreases in the cross-sectional area of the lumbar multifidus in the postures studied when compared to patients who did not have lower back pain. This idea suggests that neurological function in lower back pain is not the same as in healthy subjects during postural changes.[44] Hides et al[48] showed via real-time ultrasound that patients who had lower back pain consistently exhibited multifidus atrophy on the side of their symptoms. Therefore, chronic lower back pain patients will not show the same characteristic firing patterns of that of normal subjects and will not have the spinal stability needed during daily activities. An **intact neuromuscular system** is needed to support dynamic changes in posture, and this is exactly what the patient with lower back pain lacks.[46,47,49]

Clinical Pearls: Multifidus

- Primary function is to aid in extension
- Will aid in lateral flexion
- Will aid in axial rotation

- Helps to minimize spine stress during trunk flexion

- Will work with the abdominal muscles to balance the spine during activities, especially the transverses muscle

References

1. Adams MA, Dolan P, Hutton WC. Diurnal variations in the stresses on the lumbar spine. *Spine.* 1987;12(2):130-137.
2. Hollinshead WH, Rosse C. *Textbook of Anatomy.* 4th ed. Philadelphia: Harper & Row; 1985.
3. Clemente CD. *Gray's Anatomy.* 30th ed. Philadelphia: Lea and Febiger; 1985.
4. Adams MA, McMillan DW, Green TP, Dolan P. Sustained loading generates stress concentrations in lumbar intervertebral discs. *Spine.* 1996;21(4):434-438.
5. Panjabi MM. Clinical spinal instability and low back pain. *J Electromyogr Kinesiol.* 2003;13(4):371-379.
6. Pintar FA, Yoganandan N, Myers T, Elhagediab A, Sances A Jr. Biomechanical properties of human lumbar spine ligaments. *J Biomech.* 1992;25(11):1351-1356.
7. Johnson GM, Zhang M. Regional differences within the human supraspinous and interspinous ligaments: a sheet plastination study. *Eur Spine J.* 2002;11(4):382-388.
8. Solomonow M, Zhou BH, Harris M, Lu Y, Baratta RV. The ligamento-muscular stabilizing system of the spine. *Spine.* 1998;23(23):2552-2562.
9. Panjabi M, Abumi K, Duranceau J, Oxland T. Spinal stability and intersegmental muscle forces: a biomechanical model. *Spine.* 1989;14(2):194-200.
10. Shirazi-Adl A. Analysis of large compression loads on lumbar spine in flexion and in torsion using a novel wrapping element. *J Biomechanics.* 2006;39(2):267-275.
11. Panjabi M, Chang D, et al. An analysis of errors in kinematic parameters associated with in vivo functional radiographs. *Spine* (Phila Pa 1976). 1992;17(2):200-205.
12. Yucesoy CA, Koopman BH, Baan GC, Grootenboer HJ, Huijing PA. Extramuscular myofascial force transmission: Experiments and finite element modeling. *Arch Physiol Biochem.* 2003;111(4):377-388.
13. Quint U, Wilke HJ, Shirazi-Adl A, Parnianpour M, Loer F, Claes LE. Importance of the intersegmental trunk muscles for the stability of the lumbar spine: a biomechanical study *in vitro. Spine.* 1998;23(18):1937-1945.
14. Warfel J. *The Extremities: Muscles and Motor Points.* Philadelphia: Lea and Febiger; 1985.
15. Standring S. *Gray's Anatomy: The Anatomical Basis of Clinical Practice.* 40th ed. New York: Churchill Livingstone; 2008.
16. Indahl A, Kaigle AM, Reikeras O, Holm SH. Interaction between the porcine lumbar intervertebral disc, zygapophysial joints, and paraspinal muscles. *Spine.* 1997;22(24):2834-2840.
17. Kasra M, Shirazi-Adl A, Drouin G. Dynamics of human lumbar intervertebral joints: experimental and finite-element investigations. *Spine.* 1992;17(1):93-102.
18. Hides JA, Jull GA, et al. Long-term effects of specific stabilizing exercises for first-episode low back pain. *Spine.* 2001;26(11):E243-E248.
19. McGill SM. Low back stability: from formal description to issues for performance and rehabilitation. *Exerc Sport Sci Rev.* 2001;29(1):26-31.
20. Richardson CA, Snijders CJ, Hides JA, Damen L, Pas MS, Storm J. The relation between the transversus abdominis muscles, sacroiliac joint mechanics, and low back pain. *Spine.* 2002;27(4):399-405.

21. Stokes IA, Gardner-Morse M. Spinal stiffness increases with axial load: another stabilizing consequence of muscle action. *J Electromyogr Kinesiol.* 2003;13(4):397-402.

22. Bergmark A. Stability of the lumbar spine: a study in mechanical engineering. *Acta Orthop Scand Suppl.* 1989;230:1-54.

23. Granata KP, Marras WS. Cost-benefit of muscle co-contraction in protecting against spinal instability. *Spine.* 2000;25(11):1398-1404.

24. Holm S, Indahl A, Solomonow M. Sensorimotor control of the spine. *J Electromyogr Kinesiol.* 2002;12(3):219-234.

25. Sapsford RR, Hodges PW, Richardson CA, Cooper DH, Markwell SJ, Jull GA. Co-activation of the abdominal and pelvic floor muscles during voluntary exercises. *Neurourol Urodyn.* 2001;20(1):31-42.

26. Mens JM, Vleeming A, Snijders CJ, Koes BW, Stam HJ. Validity of the active straight leg raise test for measuring disease severity in patients with posterior pelvic pain after pregnancy. *Spine.* 2002;27(2):196-200.

27. Lee D. *The Pelvic Girdle: An Approach to Examination and Treatment of the Lumbo-Pelvic-Hip Region.* 2nd ed. New York: Churchill Livingstone; 1999.

28. Moseley GL, Hodges PW, Gandevia SC. Deep and superficial fibers of the lumbar multifidus muscle are differentially active during voluntary arm movements. *Spine.* 2002;27(2):E29-E36.

29. Wilke HJ, Wolf S, Claes LE, Arand M, Wiesend A. Stability increase of the lumbar spine with different muscle groups: a biomechanical *in vitro* study. *Spine.* 1995;20(2):192-198.

30. Meadows J. *Differential Diagnosis in Physical Therapy: A Case Study Approach.* New York: McGraw-Hill; 1999.

31. Meadows J. University of Baltimore, Physical Therapy Continuing Education Classes, 2005. Baltimore, MD.

32. Bogduk N. *Clinical Anatomy of the Lumbar Spine and Sacrum.* 4th ed. London: Churchill Livingstone; 2005.

33. Hides JA, Richardson CA, et al. Multifidus muscle recovery is not automatic after resolution of acute, first-episode low back pain. *Spine* (Phila Pa 1976). 1996;21(23):2763-2769.

34. MacDonald DA, Moseley GL, et al. The lumbar multifidus: does the evidence support clinical beliefs? *Man Ther.* 2006;11(4):254-263.

35. Wu PB, Date ES, et al. The lumbar multifidus muscle in polysegmentally innervated. *Electromyogr Clin Neurophysiol.* 2000;40(8):483-485.

36. Donisch EW, Basmajian JV. Electromyography of deep back muscles in man. *Am J Anat.* 1972;133(1):25-36.

37. Bogduk N, Macintosh JE, et al. A universal model of the lumbar back muscles in the upright position. *Spine* (Phila Pa 1976). 1992;17(8):897-913.

38. Lewin T, Moffett B. The morphology of the lumbar synovial interveertebral joints. *Acta Morphol Neerl Scand.* 1962;4:299-319.

39. Macintosh JE, Bogduk N. Volvo award in basic science. The morphology of the lumbar erector spinae. *Spine* (Phila Pa 1976) 1987;12(7):658-668.

40. Lee SW, Chan CK, Lam TS et al. Relationship between low back pain and lumbar multifidus size at different postures. *Spine* (Phila Pa 1976). 2006;31(19):2258-2262.

41. Coste J, Lefrancois G, et al. Prognosis and quality of life in patients with acute low back pain: insights from a comprehensive inception cohort study. *Arthritis Rheum.* 2004;51(2):168-176.

42. Jenkins EM, Borenstein DG. Exercise for the low back pain patient. *Baillieres Clin Rheumatol.* 1994;8(1):191-197.

43. MacDonald D, Moseley GL, et al. Why do some patients keep hurting their back? Evidence of ongoing back muscle dysfunction during remission from recurrent back pain. *Pain.* 2009;142(3):183-188.

44. Hides JA, Stokes MJ, Saide M, Jull GA, Cooper DH. Evidence of lumbar multifidus muscle wasting ipsilateral to symptoms in patients with acute/subacute low back pain. *Spine.* 1994;19(2):165-172.

45. Hides J, Gilmore C. Multifidus size and symmetry among chronic LBP and healthy asymptomatic subjects. *Man Ther.* 2008;13(1):43-49.

46. Ulrich Q, Wilke Hans-Joachim, Shirazi-Adl Aboulfazl. Importance of the intersegmental trunk muscles for the stability of the lumbar spine: a biomechanical study in vitro. *Spine.* 1998;23;18:15 1937-1945.

47. Lee SW, CK Chan, Lam TS, et al. Relationship between low back pain and lumbar multifidus size at different postures. *Spine* (Phila Pa 1976). 2006;31(19):2258-2262.

48. Hides JA, WR. Stanton. Effect of stabilization training on multifidus muscle cross-sectional area among young elite cricketers with low back pain. *J Orthop Sports Phys Ther.* 2008 38;3:101-108

49. Cholewicki J, McGill SM. Mechanical stability of the in vivo lumbar spine: implications for injury and chronic low back pain. *Clin Biomech (Bristol, Avon).* 1996;11(1):1-15.

Bibliography

Adams P, Eyre DR, Muir H. Biochemical aspects of development and ageing of human lumbar intervertebral discs. *Rheumatol Rehabil.* 1977;16(1):22-29.

Chiradejnant A, Maher CG, Latimer J. Objective manual assessment of lumbar posteroanterior stiffness is now possible. *J Manipulative Physiol Ther.* 2003;26(1):34-39.

Cresswell AG, Oddsson L, Thorstensson A. The influence of sudden perturbations on trunk muscle activity and intra-abdominal pressure while standing. *Exp Brain Res.* 1994. 98(2):336-341.

Dunlop RB, Adams MA, Hutton WC. Disc space narrowing and the lumbar facet joints. *J Bone Joint Surg Br.* 1984;66(5):706-710.

Ebara S, Iatridis JC, Setton LA, Foster RJ, Mow VC, Weidenbaum M. Tensile properties of nondegenerate human lumbar anulus fibrosus. *Spine.* 1996;21(4):452-461.

Hodges PW, Heijnen I, Gandevia SC. Postural activity of the diaphragm is reduced in humans when respiratory demand increases. *J Physiol.* 2001;537(Pt 3):999-1008.

Hodges PW, Richardson CA. Contraction of the abdominal muscles associated with movement of the lower limb. *Phys Ther.* 1997;77(2):132-142.

Hodges PW, Richardson CA. Delayed postural contraction of transversus abdominis in low back pain associated with movement of the lower limb. *J Spinal Disord.* 1998;11(1):46-56.

Hungerford B, Gilleard W, Hodges P. Evidence of altered lumbopelvic muscle recruitment in the presence of sacroiliac joint pain. *Spine.* 2003;28(14):1593-1600.

Kiesel KB, T Uhl, et al. Rehabilitative ultrasound measurement of select trunk muscle activation during induced pain. *Man Ther.* 2008;13(2):132-138.

Koumantakis GA, Watson PJ, Oldham JA. Supplementation of general endurance exercise with stabilisation training versus general exercise only. Physiological and functional outcomes of a randomised controlled trial of patients with recurrent low back pain. *Clin Biomech (Bristol, Avon).* 2005;20(5):474-482.

Maher C, Adams R. Reliability of pain and stiffness assessments in clinical manual lumbar spine examination. *Phys Ther.* 1994;74(9):801-809; discussion 809-811.

Maher CG, Latimer J, Adams R. An investigation of the reliability and validity of postero-anterior spinal stiffness judgments made using a reference-based protocol. *Phys Ther.* 1998;78(8):829-837.

Moseley GL, Hodges PW, Gandevia SC. External perturbation of the trunk in standing humans differentially activates components of the medial back muscles. *J Physiol.* 2003;547(Pt 2):581-587.

Pressler JF, Heiss DG, et al. Between-day repeatability and symmetry of multifidus cross-sectional area measured using ultrasound imaging. *J Orthop Sports Phys Ther.* 2006;36(1):10-18.

Rajasekaran S, Venkatadass K, Naresh Babu J, Ganesh K, Shetty AP. Pharmacological enhancement of disc diffusion and differentiation of healthy, ageing and degenerated discs: results from in-vivo serial post-contrast MRI studies in 365 human lumbar discs. *Eur Spine J.* 2008;17(5):626-643.

Sairyo K, Biyani A, et al. Pathomechanism of ligamentum flavum hypertrophy: a multidisciplinary investigation based on clinical, biomechanical, histologic, and biologic assessments. *Spine* (Phila Pa 1976). 2005;30(23):2649-2656.

Schendel MJ, Wood KB, Buttermann GR, Lewis JL, Ogilvie JW. Experimental measurement of ligament force, facet force, and segment motion in the human lumbar spine. *J Biomech.* 1993;26(4-5):427-438.

Stokes IA, Henry SM, Single RM. Surface EMG electrodes do not accurately record from lumbar multifidus muscles. *Clinical Biomechanics (Bristol, Avon).* 2003;18(1):9-13.

Strauss SE, Richardson SW, Glasziou P, Haynes RB. *Evidence-Based Medicine.* 3rd ed. London: Churchill Livingstone; 2005.

Tsao H, Hodges PW. Persistence of improvements in postural strategies following motor control training in people with recurrent low back pain. *J Electromyogr Kinesiol.* 2008;18(4):559-567.

Urquhart DM, Barker PJ, Hodges PW, Story IH, Briggs CA. Regional morphology of the transversus abdominis and obliquus internus and externus abdominis muscles. *Clin Biomech (Bristol, Avon).* 2005;20(3):233-241.

Wallwork TL, Hides JA, et al. Intrarater and interrater reliability of assessment of lumbar multifidus muscle thickness using rehabilitative ultrasound imaging. *J Orthop Sports Phys Ther.* 2007;37(10):608-612.

Wallwork TL, Stanton WR, et al. The effect of chronic low back pain on size and contraction of the lumbar multifidus muscle. *Man Ther.* 2009;14(5):496-500.

Whalen JL, Parke WW, Mazur JM, Stauffer ES. The intrinsic vasculature of developing vertebral end plates and its nutritive significance to the intervertebral discs. *J Pediatr Orthop.* 1985;5(4):403-410.

12

THE LUMBAR SPINE MUSCLES AND THEIR FUNCTION

There are 3 sets of muscles in the lumbar spine that are of importance and we have already discussed the most medial of this group: the lumbar multifidus. In this chapter, we will discuss the other supporting lumbar spine muscles. They also play a role in movement patterning and stability and are called the iliocostalis lumborum, the longissimus, and the spinales (Figures 12-1A and B). They are usually referred to as the erector spinae.[1-3]

Function

The iliocostalis lumborum causes the spinal column to extend when both left and right portions of the muscle are being contracted.[1] When a unilateral contraction is being performed, the laterally placed iliocostalis will cause lateral flexion and axial rotation toward the same side.[1,3] The longissimus is not well differentiated in the lumbar spine from the iliocostalis, but is generally intermediate between the iliocostalis and spinalis and has a vertical orientation. The spinalis (sacrospinalis) is the most medial of the erector spinae muscles[3] and forms a continuous myofascial connection between the sacrum and the lumbar spine, thoracic spine, cervical spine, and occiput (Figure 12-1C). In the lumbar and thoracic spine, the spinalis is covered by the thoracolumbar fascia. The function of the spinalis is to assist with extension of the vertebral column and also will aid in ipsilateral lateral flexion and rotation.

Origin and Insertion

Origin of the Iliocostalis Lumborum, Longissimus, and the Spinales	Insertion of the Iliocostalis Lumborum, Longissimus, and the Spinales
The iliocostalis lumborum arises from the muscle mass of the erector spinalis.	The iliocostalis lumborum will ascend upward and attach to the sixth and seventh ribs at their lower border. These 2 muscles are intimately related since the erector spinalis will give rise to the iliocostalis lumborum (see Figure 12-1C).
The longissimus in the lumbar region arises from the transverse processes and accessory processes of the lumbar vertebrae, as well as from the underside of the thoracolumbar fascia.	The longissimus will attach itself to the transverse and spinous process of the lumbar vertebrae.

(Continued)

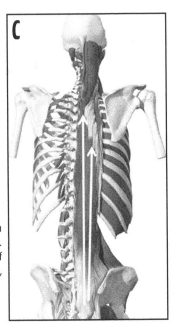

Figure 12-1. (A) Relaxed longissimus and spinalis muscles in prone. (B) The contracted longissimus and spinalis in prone. (C) Schematic of the longissimus and spinalis with arrows of force shown. (Reprinted with permission from Primal Pictures, 2009.)

Origin of the Iliocostalis Lumborum, Longissimus, and the Spinales	Insertion of the Iliocostalis Lumborum, Longissimus, and the Spinales
The spinalis arise from the sacrum, iliac crest the lumbar spinous processes and the eleventh and twelfth thoracic spinous processes (see Figures 12-1A and C).[3]	The muscle traverses upward throughout the entire vertebral column and is thinnest in the sacral region and thickest in the lumber region. Once the muscle leaves the lumbar region, it subdivides into 3 vertical columns that attach to the vertebral bodies and the ribs as they ascend (see Figures 12-B and C).

Action of the Lumbar Spine Muscles
The deep lumbar spine muscles will primarily be involved in spinal extension, lateral flexion, and axial rotation. The deep spine muscles will also aide in pelvis side motions.[1]

Pathomechanics of Common Injuries

The spinalis and the iliocostalis will cause extension, lateral flexion, and rotation as they act through long lever arms.[4] These global muscles require dynamic support from the deep fibers of the multifidus to allow movement with segmental control. As we have previously discussed, if the multifidus is not working properly, the spine will not have the necessary inherent dynamic support, increasing the risk for injury.[5-7]

Clinical Pearls

- Will aid in lumbar extension, flexion, and rotation
- Will aid in pelvic and ligament support to the lumbar spine
- Will aid in depressing the ribs and help with supporting the rib cage, thus helping with breathing

References

1. Warfel J. *The Extremities: Muscles and Motor Points.* Philadelphia, PA: Lea and Febiger; 1985.
2. Hollinshead WH, Rosse C. *Textbook of Anatomy.* 4th ed. Philadelphia, PA: Harper & Row; 1985.
3. Clemente CD. *Gray's Anatomy.* 30th ed.Philadelphia, PA: Lea and Febiger; 1985.
4. Hides JA, Richardson CA, et al. Multifidus muscle recovery is not automatic after resolution of acute, first-episode low back pain. *Spine* (Phila Pa 1976). 1996;21(23):2763-2769.
5. Hides JA, Jull GA, et al. Long-term effects of specific stabilizing exercises for first-episode low back pain. *Spine.* 2001;26(11):E243-E248.
6. Lee SW, Chan CK, et al. Relationship between low back pain and lumbar multifidus size at different postures. *Spine* (Phila Pa 1976). 2006;31(19):2258-2262.
7. Reeves NP, Cholewicki J, et al. Muscle activation imbalance and low-back injury in varsity athletes. *J Electromyogr Kinesiol.* 2006;16(3):264-272.

Bibliography

Bogduk N. *Clinical Anatomy of the Lumbar Spine and Sacrum.* 4th ed. London: Churchill Livingstone; 2005.

13

CORE MUSCLE FUNCTION

The concept of a "core" group of muscles is a very old idea that is reflected in various approaches to physical health, including yoga and, more recently, Pilates exercise. It remains to be seen whether the core muscles represent a distinct subset of the "local" muscles as defined by Bergmann or if they will be classified differently. Only in the past decade have scientific studies confirmed the existence of muscle groups that seem to be capable of stabilizing the spine while minimizing compressive forces. These deeply placed muscles have been difficult to study, especially since surface EMG has shown to be inadequate for characterizing their performance.[1] This chapter will outline the core muscle groups and functionality of this muscle system.

Function of the Core Muscles

The muscles in this core group include the transversus abdominis (TrA), multifidus, diaphragm, and pelvic floor muscles (Figures 13-1A and B). These muscles appear to have a substantial role in maintaining spine stability as well as additional roles in producing spinal movement, respiration, and other bodily functions.[2,3]

The TrA shows some regional variations in fiber direction and septation (ie, the number of different fiber directions) that may reflect its multiple functions (see Figures 13-1A and B).[4] As we turn to the lumbar multifidus, there is evidence to show that the muscle is functionally divided into superficial and deep fibers, which respond differentially to movements. It is the deep fibers of the multifidus that appear to mirror the stabilizing function of the transversus abdominis, since their activation is insensitive to movement direction (Figure 13-2A).[5] The basic idea of the core muscles is to be tonically to pre-emptively defends the spine as dynamic stabilizers.

Research led by a team of Australian scientists has provided a steady stream of discoveries about the function of these muscles over the past decade.[6-9] Findings include the fact that electrical activity in the healthy TrA and other core muscles typically precedes voluntary arm movements by several hundred milliseconds, and that this pre-emptive activity disappears or is significantly delayed in the presence of back pain. Thus, if a patient experiences lower back pain, there will be a delay in the TrA firing and support is reduced.

Figure 13-1A. The abdominal muscles relaxed in supine.

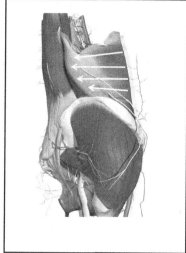

Figure 13-1B. Internal transversus abdominis with the origin shown with the star and the arrows showing the forces from the insertion to the origin. (Reprinted with permission from Primal Pictures, 2009.)

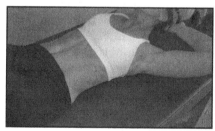

Figure 13-2A. The contraction of the pelvic floor muscles and the abdominals in supine with the transversus abdominis contracted.

Figure 13-2B. Schematic of the pelvic floor muscle in the female pelvis (frontal view) and arrows showing the forces from insertion to origin. (Reprinted with permission from Primal Pictures, 2009.)

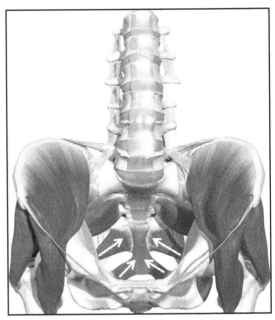

These research findings have been incorporated into practice by clinicians in many disciplines, and a recent study supports this "motor control" approach to lower back pain[10] with the aim of restoring normal pre-emptive function to the core muscles. Of course, back pain and back surgery are complicated issues, and focusing exclusively on the core muscles alone will not restore a patient's spine to healthy function. Focusing on the core systems first, then incorporating the other low back and hip muscles as well as aerobic conditioning, will give health professionals a good working approach to helping the patient with low back pain avoid chronic lower back pain.

Origin and Insertion

Origin of the Core Muscles (Transversus, Diaphragm, and Pelvic Floor)	Insertion of the Core Muscles (Transversus, Diaphragm, and Pelvic Floor)
The transversus abdominis muscle is the deepest abdominal muscle. It arises from the cartilage of the lower 6 ribs, the iliac crest, the thoracolumbar fascia, and the inguinal ligament (see Figure 13-1B).[11]	The fibers run transverse to the linea alba and the rectus abdominis, circling the waist like a corset. The lower fibers of the TrA will run obliquely downward toward the pubic bone. The majority of fibers will run with the internal oblique and will attach to the linea alba (see Figure 13-1B).[11,12]

(Continued)

DIAPHRAGM	
The sternocostal fibers of the diaphragm come from the inner portion of the xyphoid process (sternal portion) and the inner side of the lower 6 ribs (costal portion). The lumbar portion of the muscle arises from the anterior surfaces of the upper 2 or 3 lumbar vertebra.	The 3 portions of the diaphragm converge into a central tendon that contains a posterior hiatus through which the esophagus and great vessels pass into the abdominal cavity.[11]
PELVIC FLOOR MUSCLES (LEVATOR ANI) *Iliococcygeus:* This muscle is a broad, thin muscle that forms the posterior portion of the pelvic floor. The posterior fibers arise for the ischial spine and the other fibers will arise from the obturator fascia (see Figure 13-2B).	The iliococcygeus muscle fibers travel transversely to attach to the side and tip of the posterior portion of the coccyx. The more anterior fibers of the iliococcygeus also attach to the coccyx, meeting in the midline near the anus.[11]
Pubococcygeus: This muscle is the more anterior portion of the levator ani and arises from the posterior portion of the pubic bone and the anterior portion of the obturator fascia (see Figure 13-2B).[11]	The pubococcygeus muscle fibers will traverse backwards to attach to the coccyx and sacrum (see Figure 13-2B).[11]
Action of the Core Muscles: The core muscles are designed to provide stability to the spine and pelvis with their ability to "draw in" and this is the effect of the TrA and the pelvic floor muscles. The diaphragm is a major muscle for respiration that also will have a direct effect on lower back functioning.[9]	

Clinical Pearls: Core Muscles

- Help to support the sacroiliac joint
- Aid with respiration
- Support the pelvic floor and aid in trunk support
- Work together during activities
- Aid in maintaining continence and are more important in females then males
- Are active during lifting and recreational activities
- Are needed to work in concert to reduce the likelihood of low back pain

References

1. Stokes IA, Gardner-Morse M. Spinal stiffness increases with axial load: another stabilizing consequence of muscle action. *J Electromyogr Kinesiol.* 2003;13(4):397-402.

2. Hodges PW, Heijnen I, Gandevia SC. Postural activity of the diaphragm is reduced in humans when respiratory demand increases. *J Physiol.* 2001;537(Pt 3):999-1008.

3. Moseley GL, Hodges PW, Gandevia SC. External perturbation of the trunk in standing humans differentially activates components of the medial back muscles. *J Physiol* 2003;547(Pt 2):581-587.

4. Urquhart DM, Barker PJ, Hodges PW, Story IH, Briggs CA. Regional morphology of the transversus abdominis and obliquus internus and externus abdominis muscles. *Clin Biomech (Bristol, Avon).* 2005;20(3):233-241.

5. Moseley GL, Hodges PW, Gandevia SC. Deep and superficial fibers of the lumbar multifidus muscle are differentially active during voluntary arm movements. *Spine.* 2002;27(2):E29-E36.

6. Cresswell AG, Oddsson L, Thorstensson A. The influence of sudden perturbations on trunk muscle activity and intra-abdominal pressure while standing. *Exp Brain Res.* 1994;98(2):336-341.

7. Hides JA, Stokes MJ, Saide M, Jull GA, Cooper DH. Evidence of lumbar multifidus muscle wasting ipsilateral to symptoms in patients with acute/subacute low back pain. *Spine.* 1994;19(2):165-172.

8. Hodges PW, Richardson CA. Contraction of the abdominal muscles associated with movement of the lower limb. *Phys Ther.* 1997;77(2):132-142.

9. Hodges PW, Richardson CA. Delayed postural contraction of transversus abdominis in low back pain associated with movement of the lower limb. *J Spinal Disord.* 1998;11(1):46-56.

10. Tsao H, Hodges PW. Persistence of improvements in postural strategies following motor control training in people with recurrent low back pain. *J Electromyogr Kinesiol.* 2008;18(4):559-567.

11. Hollinshead WH, Rosse C. *Textbook of Anatomy.* 4th ed. Philadelphia: Harper & Row; 1985.

12. Clemente CD. *Gray's Anatomy.* 30th ed. Philadelphia: Lea and Febiger; 1985.

Bibliography

Bergmark A. Stability of the lumbar spine: a study in mechanical engineering. *Acta Orthop Scand Suppl.* 1989;230:1-54.

Bogduk N. *Clinical Anatomy of the Lumbar Spine and Sacrum.* 4th ed. London: Churchill Livingstone; 2005.

Bogduk N, Macintosh JE, et al. A universal model of the lumbar back muscles in the upright position. *Spine* (Phila Pa 1976). 1992;17(8):897-913.

Lee D. *The Pelvic Girdle: An Approach to Examination and Treatment of the Lumbo-Pelvic-Hip Region.* 2nd ed. New York: Churchill Livingstone; 1999.

FINANCIAL DISCLOSURES

Scott Benjamin has no financial or proprietary interest in the materials presented herein.

Roy Bechtel has no financial or proprietary interest in the materials presented herein.

Vincent Conroy has no financial or proprietary interest in the materials presented herein.

INDEX

Printed in the United States
by Baker & Taylor Publisher Services